Nature by
Night

A GUIDE TO OBSERVATION AND IDENTIFICATION

TEXT
Vincent Albouy and Jean Chevallier

ILLUSTRATIONS
Jean Chevallier

NH
NEW
HOLLAND

Acknowledgements:
Jean Chevallier would like to thank C. Bretagnolle,
E. Chapoulie, M. Jay and G. Lesaffre for their help and
advice, as well as all the friends who have joined him
on night-time wildlife-watching trips, for all those
shared experiences.

Publishing Director: Rosemary Wilkinson
Commissioning Editor: Simon Papps
Translation: Alison Taylor-Granie

ISBN 978 1 84773 114 2

This edition first published in 2008 by
New Holland Publishers (UK) Ltd
Garfield House, 86-88 Edgware Road, London W2 2EA
www.newhollandpublishers.com

10 9 8 7 6 5 4 3 2 1

Title of the original edition: *La nature la nuit*
published by Delachaux et Niestlé, Paris

Text: Vincent Albouy, except for mammal and bird
entries by Jean Chevallier
Illustrations: Jean Chevallier
Design: Nicolas Hubert

Printed and bound in Slovenia by MKT Print

Contents

I

GENERAL

INFORMATION

The night: another world

Why watch wildlife at night?

The average human spends about one third of their lifetime sleeping. That is eight hours per day, or more commonly per night. This natural cycle has only recently been disturbed by the abundance of artificial light generated by low-cost electricity. Like a lot of other species, we sleep at night because we are adapted to diurnal life. When vision is the primary sense for appreciating the immediate environment, darkness tends to bring the bulk of activity to a halt. However, there are many species which have evolved so that they are predominantly active at night, choosing instead to rest and shelter during the hours of daylight.

This book will take you on a voyage of discovery into this other world, the night, which is so similar yet so different to the daytime world that we know best. It cannot instantly make you an expert on all the techniques for studying wildlife at night – these can only be honed through personal experience coupled with a good background knowledge of nature. It does, however, present an introduction to the fascinating world of the night, offering an approach to watching nocturnal wildlife and the basic knowledge necessary to stand a good chance of coming home from an outing having seen something truly memorable.

For humans, the night is characterised by the absence of light from the sun. Even if other natural light sources exist, the most important one being the moon, at best they represent just 7 per cent of the sun's luminosity. At night our eyes, adapted to the abundance of the sun's rays, lose their capacity to perceive colour. Even with the most beautiful full moon, we can only distinguish our environment in black and white. If it is very dark, for example when the sky is overcast, we literally lose our bearings and become incapable of moving without having to feel our way around.

Many species have adapted to this rare or absent light. Some, such as owls or genets, have distinctly improved the sensitivity of their eyesight, enabling them to takeg advantage of simple starlight. Others, or sometimes the same species, have developed alternative senses. Hearing is accentuated for mice and owls and reaches near perfection in bats with their ultrasound echolocation. Sense of smell is highly developed for Hedgehogs and moths, and that of touch for shrews or cockroaches.

A variety of animals, like foxes or dung-beetles, have both diurnal and nocturnal habits. If they prefer to be active at night, it is to protect themselves from daytime disturbance caused by humans or other potential predators. Deer tend to call at night, because the same behaviour during the day would dangerously expose them to predators. Likewise many birds, particularly passerines, choose to migrate at night.

Many species have definitively crossed the barrier and are only active at night. To escape the numerous diurnal insectivores, the vast majority of moth species have adapted to nocturnal life. Their dull colours and cryptic patterns render them almost invisible during the day as they remain immobile while roosting. This enables them to escape from an army of insectivores, especially birds.

As moths adapted, so did their predators. Bats and birds such as nightjars have evolved to tap into this abundant food supply. In fact, entire nocturnal ecosystems exist. For example, certain flowers ly open to secrete their perfume and nectar at night so that they can take advantage of the efficient pollinating ability of moths.

Going unnoticed by predators is not the only advantage of nocturnal life. The disappearance of the sun and the heat that it provides brings about a temperature drop, which in turn causes an increase in the air's humidity. This is why many species which have difficulty maintaining their internal body temperature, or which are sensitive to the dryness of the air, find that conditions are more suitable if they are active by night.

Amphibians, worms, slugs, snails and woodlice fall into this category. Light is not directly involved in this behaviour, which is why it is common to see these animals during a stormy, cloudy summer's afternoon, when the temperature has dropped and air humidity has increased.

Wild Boar feeding by moonlight

Equipment and clothing

Observing nature at night does not necessarily require a great deal of specialist equipment and clothing. Clothing should be comfortable, warm and waterproof if there is a chance of rain, and it should not be noisy or restrict movement. Clothes should preferably be dark and matt to avoid being visible as a patch of light on a dark background in the moonlight. It is particularly important that shoes support the ankles, as there is a greater risk of sprains when you cannot see the ground underfoot. Also make sure that they are waterproof, as dew can quickly make for wet feet and a miserable trip.

A rucksack can be useful for storing a jumper, waterproof, drink, snack, torch, guidebook or other materials that are too cumbersome to carry while walking. A strip of foam rubber such as a camping mat is also a good idea as it allows you to lie down more comfortably on rough ground or to sit on rocks during an extended vigil.

When hoping to watch wary creatures with good nocturnal vision, such as mammals or birds, it is a good idea to mask white patches in any way possible. This can include making sure that the face and hands are fully covered.

At dusk and dawn, and sometimes by moonlight, light levels can be sufficiently bright to enable the use of binoculars. The low light intensity means that certain specifications perform better than others. Fully multicoated optics transmit light to the eye more effectively, while models with larger objective lenses tend to give a brighter image – 8x40 and 10x50 binoculars are well adapted to viewing in low light levels.

Small fauna and flora can only be observed at night with the help

A torch-beam reveals a Bittern at a reedbed roost site

of a torch. If a torch is to be used regularly, it is wise to invest in an efficient model. Headlamps, which free your hands and automatically direct the light beam where you are looking, are the most practical. Using rechargeable batteries will considerably reduce cost and dangerous waste.

Understanding the night

The moon illuminates a
December night

Although humans are not accustomed to being active at night,
our species is characterised by its adaptability. We can, to a certain
extent, modify our perceptions and senses in order to partially make
up for the absence of light.

We can still use our eyes as it is rarely pitch-black at night. Our
pupil dilates or retracts according to brightness. If there is too much
light its diameter decreases to limit the entry of luminous rays to the
retina. Conversely, if the amount of light decreases it opens up to a
maximum of 7 mm in young adults, who have the sharpest vision.
Performance deteriorates with age.

It takes about 15 minutes for the pupil to dilate to its maximum in
the dark and roughly half an hour more for the retina to be perfectly
adjusted to these very low light conditions. Therefore, it is a good
idea to spend 30 to 45 minutes in the dark before the start of a trip in
order to be functioning fully from the outset.

The use of a torch, even just short flashes, hinders this adaptation,
and the resulting dazzle reduces the eyes' performance for several
minutes or more. Switching on and off is therefore counterproductive.
It is better to decide beforehand whether the outing will be with a

torch (for observing plants, invertebrates or amphibians) or without (for birds and mammals).

Immersion in darkness refines the senses and brings to the fore those which we consider to be less important. After sight, our most important sense is hearing. When the eyes are no longer capable of sending sufficient information about the environment to the brain, the brain compensates by soliciting the ears and we become more sensitive to noises around us.

We are able to distinguish background noise (the wind in the trees or busy roads) from more unusual sounds which can have a real informative value. At the same time, known or common noises, such as a dog barking, a car engine or a passing train, make less of an impression than an unknown noise, such as the call of a wild animal or bird.

Weather conditions can increase or decrease the distance of sound transmission, or distort the sound itself. The soundscape in which we are immersed is constantly changing and is very different to the visual landscape: it is a different way of perceiving the same reality.

If hearing becomes our most important sense in the dark, it is more difficult to summon our other two senses, touch and smell, which are called upon the least. Touch only allows us to familiarize ourselves with our immediate environment. It only plays a secondary role in the discovery of the world of the night. However, in the case of total darkness, we spontaneously rely on touch, holding our arms out in front of us to try and feel what is nearby.

On the other hand, smell can provide a good deal of information, but only on two conditions. Firstly, that we have a sufficiently sensitive sense of smell, which is not the case for everyone. Secondly, that we are capable of distinguishing and identifying the odours around us, which requires a little experience and a good memory.

Understanding the night is a question of practice. Do not expect to hear, smell, identify and interpret everything on your first night-time outing. Over time, though, your accumulated experience will allow your secondary senses to develop, sometimes without you even being aware of it.

The right attitude

Young children learning to play hide-and-seek go through an initial phase where they believe that they are invisible to others when they cover their eyes with their hands. We laugh at their naivety, but there is a strong risk of making the same mistake at night. Not being able to easily see our surroundings does not make us invisible, and certainly not to nocturnal creatures, many of which have much better vision than humans. Whether you are taking a random walk, going out in search of a particular species or meeting up with an organized group, the advice for watching wildlife which is valid during the day is also valid at night. For example, avoid your silhouette being visible from afar when following the crest of a hill or bank, and instead of crossing the middle of a meadow or glade, follow the shelter of an embankment or hedge. Avoid sudden gestures and adopt measured and slow movements. If you feel your presence is unsettling or alarming certain animals, stop and keep a low profile until calm is restored.

If they can see well at night, most animals can smell even more acutely. Any approach strategy which does not take wind direction into account is doomed to failure. Ideally, you should walk into the wind, or alternatively through a crosswind, so that your odour is not transmitted. This precaution only concerns smell. You should also make sure you are not visible and do not cause direct or indirect noises, especially by your movements (stones underfoot or noisy fabric, for example).

A night outing should be planned in advance. It is difficult to get one's bearings at night, when landmarks which are obvious in the light of day disappear. It is, therefore, advisable to explore an area that you already know well. In an unknown environment, a nocturnal excursion should be preceded by a few daytime scouting walks. You will appreciate the captivating atmosphere of the night and the encounters you make all the more if you don't have to worry about getting lost.

Find out about the weather forecast before you set out. Rain, snow, storms, high winds, cold and heat all have different impacts on the fauna, and therefore on what you will see and hear. This will also allow you to choose the right clothing and take adequate provisions, or even to postpone the outing. Such a decision is sometimes hard to make. It is never pleasant to experience a heavy storm in the middle of the night, but often, before and after such an event, there are moments of excitement when animals come out and are easier to approach. Activity cycles must be taken into account to avoid disappointment, especially if you are on the lookout for a particular species. Cycles vary according to species, seasons, locations and timing. Dusk and the beginning of the night, and then at the end of the night and dawn tend to be the periods of greatest activity.

The middle of the night is generally quieter, even though there may still be a good deal of activity. If you have chosen to spend the night outdoors, from dusk until dawn, this is the best time to take a short nap as it will enable you to be fresh, alert and attentive when activity intensifies as sunrise draws closer.

Successful night-time wildlife-watching depends on being in the right place at the right time. The most refined and technically perfect approach is of no use whatsoever if the creature you want to observe is not located where you are. Unless you decide to venture out with no preconceived idea of what you are going to see, you should prepare your outing according to the particular habits and behaviour of the species that you are targeting.

Practical experience from night outings, with their unexpected encounters, is the best training for recognising where and when to watch certain species. A field notebook, for jotting down both insignificant and important facts during nights under the stars, will help to aid a memory which can never remember everything nor keep it forever.

A torch beam reveals a Marbled Newt preying on an Alpine Newt. The two species occur together in western France and northern Spain

Mistakes to avoid

As smell is one of our minor senses, we tend to ignore many of the odours around us, especially when they are familiar ones. If you go out searching for birds, amphibians or invertebrates, the question is of little importance as they do not react to our odour. However, if you wish to discover mammals, neglecting this point can lead to failure.

One particular problem of group outings is what could be called the 'light bubble'. There is a tendency to talk, turn on a torch and gesture in order to show something to others. In short, acting in a manner that is contrary to that necessary for observing wildlife. Group outings do undeniably have their advantages, such as sharing sightings, knowledge and experience, but a group of three or four keen observers is the maximum that is practical if searching for birds and mammals. Above this number, the risk of disturbing the animals by the presence of the group alone becomes very high. It is therefore better for groups to concentrate on discovering flora, insects and aquatic fauna, which are much less sensitive to this kind of disturbance.

Being aware of the weather conditions is even more important at night than for daytime outings, especially in difficult terrain such as mountains, or in winter. A violent storm which brutally swells the streams and blocks the return path, or heavy snowfall or thick fog which engulf the landscape and paths are just some of the many hazards which can spoil an outing or even render it dangerous. On the other hand, calm weather and a starry night generally guarantee a good harvest of sightings, even if the creatures you were searching for did not make an appearance.

Checking the weather forecast is an essential factor in planning your trip. Nocturnal wildlife is always sensitive to the weather and many species can become hard to find in certain conditions. If the night is too cold, lower than 10-12°C, very few insects will be active. Also, high winds will prevent many winged insects from taking to the air as they will not take the risk of being carried away by a gust. Even birds abandon flight and avoid perching in exposed areas. Other factors can count towards success or failure, such as paying attention to tide times to gauge the best period to watch wading birds on the shoreline.

Of course it is hard to generalise as each situation, species and environment is unique. You will inevitably make mistakes, but you will learn a lot from these and by avoiding them in the future you will progress and refine your nocturnal observation skills.

Badgers grooming

Cautionary advice

The dangers of a night outing can be minimised by preparation and good knowledge through daytime visits to the site. This should never be neglected and the responsibility is even greater if children are present, with reinforced precautions being necessary.

Even with adequate moonlight, shadows are numerous and our visual perception of the land can be very limited. Ruts, holes and ditches are not always easy to see. Traversing difficult areas such as slopes, thickets and marshes can be very complicated. Slippery ground, an unnoticed hollow, or an optical illusion pertaining to the distance of an obstacle can all lead to a fall and ultimately to injury.

The psychological element must also be taken into consideration, especially with beginners. The night's atmosphere, with the absence of light, can generate a feeling of fear in some people, and in children in particular. This can be transformed into anxiety or even panic if there is an accident. Becoming acquainted with the night is a process that must be carried out in stages, so that its mysteries are uncovered little by little. What we understand is generally not frightening. The night outing must therefore be on a par with a hike in unfamiliar conditions. Informing a newcomer of the route, scheduled itinerary and return time is always worthwhile in order to give reassurance.

The widespread use of mobile phones offers a link to civilisation from many of the most remote spots. In order to preserve the ambience of such places, the handset really belongs in the rucksack, having first taken the precaution of saving emergency numbers into the memory, charging the battery and switching it off before leaving. Nothing is worse for spoiling a good wildlife-watching moment than the untimely ringing of a mobile. In isolated regions and in the

A winter trip to the mountains requires thorough preparation, particularly for night-time excursions

mountains, where the mobile is potentially at its most useful, do not forget that there may well be a lack of signal which will prevent all communication.

Your rucksack should always be big enough to hold a few basic necessities. Even if you don't think you will need it, a torch with batteries is essential. In the event of a problem it allows you to negotiate the terrain more easily, find you way back or if necessary find a new route. Above all, if you go out alone, in the event of a fall or serious injury preventing you from moving, the torch can at least signal your presence from a distance.

A first aid kit can treat the kind of minor injuries which occur frequently. It is particularly useful if children are present on an outing. Psychologically, the simple action of treating cuts and bruises can make the pain go away or at least soothe it.

A good knife is an all purpose tool that can be used for everything from slicing bread to cutting branches to make a hide or a rough shelter. It is an essential accessory for many naturalists and often proves to be useful in the event of unforeseen circumstances.

In places that you do not know well an accurate map and a compass will help you find your way, or at least a way, without any problems. In much of Britain and Europe, even in the mountains, human habitation is usually sufficiently dense to be able to reach an inhabited place in a few hours if you don't go round in circles.

Winter outings must be particularly well prepared for. The risk of becoming tired and weak quickly are not to be ignored in the event of an accident or getting lost.

The naturalist's code of ethics

It is great to explore the wonderful world of nocturnal wildlife but it is not a playground where satisfying the curiosity of the observer is all that matters. As a protector of nature the naturalist's priority is not to find a certain species at all costs, like a collector who wants to pin that last missing butterfly into a box, but to observe the creatures with respect.

Sometimes it is necessary to have the intelligence to turn around and abandon an outing rather than to forcefully intrude into a territory which is not ours. Failure, whether it is voluntary or involuntary, is not an uncommon result for the night-time naturalist. This makes the joy of a successful encounter all the more valuable. Any outing, even if the objective is not accomplished, serves to provide experience, memories and greater knowledge of wildlife.

Waiting for a creature to come to you rather than forcefully searching for it is a responsible attitude and one that often pays dividends. Approaching and watching an animal must be done as discreetly as possible if you want to get a good sighting – that is, sharing a moment of the animal's life during which it goes about its business unworried and unhindered. This discretion must be maintained at all times in order to cause as little disturbance as possible to the areas explored, and to avoid any collateral damage. It is not necessarily the species that we are observing or looking for that will be the most affected by disturbance.

Certain creatures are more sensitive to disturbance at particular times of year. The breeding season is one of them, especially for birds. Birds build their nest in a place which is judged to be safe and protected from predators. Disturbance by an overcurious naturalist is interpreted as a high danger signal: that the location is no longer safe. Birds will often abandon eggs or young in such situations, if their survival is judged to be uncertain in such an exposed place. At best, they will go and build another nest somewhere else and attempt a second brood, but this will have less chance of success than the first. In these conditions, it is wise to limit the length of time spent watching and waiting so as not cause irretrievable damage. Disturbance of roosting diurnal creatures can be an equally important issue. Like us, they tend to rest and sleep at night. The typical day of a bird during the spring, when it is constantly feeding its young, is exhausting. It is even worse during the winter with the intense cold and when food is scarce and difficult to find. Having a rest period is essential, and any disturbance can be fatal. Another problem is that, also like us, these species do not have good night vision. Unsettled from their comfortable, safe perch, the birds are thrown into a panic and are incapable of finding safe shelter. They risk being captured by a nocturnal predator or dying of cold as they may be isolated and badly sheltered and have to use up too much of their low energy reserves. Spells of extreme cold and snow are difficult times for many creatures, especially in places where these conditions are uncommon. Even in mountainous regions, where the animals are used to snow, its arrival makes their lives more difficult. The cold which accompanies the snow increases their energy requirements, while food which is already scarce becomes even more difficult to find. This is often illustrated by a decrease in the escape distance. Animals, which are reduced to saving their meagre strength, only flee at the last moment when the danger appears to be urgent. For them, exhaustion has become a threat as important if not more so than predation.

A cock Capercaillie slips away after being disturbed

The increased ease with which birds and mammals can be approaching during winter should be considered as an alarm signal. They are on the verge of exhaustion and tracking them too closely can only accelerate this and increase the chances of mortality. This is why hunting is forbidden in some parts of Europe during periods of snow or frost, to protect game which is incapable of reacting properly to the hunt. The naturalist who respects the wildlife he or she wishes to observe must also abstain from tracking them during these difficult times, or at least stay far enough away so as not to reduce their already decreased odds of survival.

Watching

Dusk and dawn

The transition period between day and night, between light and darkness, dusk and dawn are times which enable the naturalist to encounter a fascinating mixture of diurnal and nocturnal animals. In fact many species are crepuscular, meaning that they are most active at these times. If you are planning an encounter with nightlife on unfamiliar ground, start with these two periods which are generally the liveliest.

If there is some degree of light we are able to observe with our main sense, sight, without any specialist equipment. A good pair of binoculars is easily used and considerably increases the observer's viewing ability and range if situated in a spot with a wide perspective. This is also the optimum time for photographers, who can target bats coming out of their shelter one by one, a Badger trotting towards its set, fox cubs playing in front of their den or an owl watching from its favourite perch.

Dusk is a good time for walking, observing and discovering night flora. Visible to the naked eye, the opening of the evening-primrose, the white spots of the campion's corolla in the falling darkness and the fragrance of honeysuckle or Dame's-violet can all be appreciated at nightfall. During the spring, dusk is a good time for listening to certain birdsongs, such as that of the Nightingale. Conditions are also ideal for encountering insects. In the warm atmosphere, cicadas, crickets and grasshoppers sing their monotonous songs. Gradually, as the night progresses and the nocturnal cold takes hold, the singers quieten one by one. It is only on exceptionally hot nights that the insects sing until morning.

Sunset signals the appearance of numerous species of moths which come gathering nectar from flowers. In the garden, phlox, petunia, lilac, buddleia, four-o'clock and Dame's-violet all prove especially popular. Among the wild flora, campion, Soapwort, geranium, willow (late winter), Sweet Chestnut (late spring) and Ivy (autumn) are all attractive to nectar-gatherers at dusk.

If you come across a hawk-moth stationary on a branch or trunk during the day, go back to witness its awakening in the evening. When darkness begins to fall, the insect trembles, stretches out its antenna, splays its wings and vibrates them to warm up its muscles. When the engine is warm, it majestically takes flight, turns around once or twice and heads off to fill up with nectar.

In the cool conditions of evening, creatures which are incapable of enduring the dryness of summer days, such as snails, slugs, woodlice and toads, come out of their shelters and start searching for food. On stormy, humid days with little light they often emerge in late afternoon if the conditions they require are all present.

Dawn can be a more advantageous time for watching. A hide can be reached in darkness with less chance of drawing attention to yourself. The end of the night becomes animated with the last actions of marauding predators such as a Hedgehog or a fox, and also with a group of Roe Deer at the edge of a wood or hoard of Wild Boar rooting around a pasture with their snouts.

At the end of winter and the beginning of spring a chorus of birdsong begins to build. For those with a good ear and who have learnt to recognise the songs of different species, this is a good way of detecting which songbirds are present at a particular site.

Due to the drop in temperature, the end of the night and dawn are times when insects are less active than at dusk. On both garden vegetables and wild vegetation, this is the best time to find caterpillars devouring leaves – they often hide on the ground or in litter during the day, going unnoticed by the eyes of the diurnal observer.

When the light of dawn is sufficiently bright, explore the tops of grass stalks and flowers. You will see and be able to photograph numerous stationary diurnal insects, especially nectar-gathering bumblebees and butterflies. During the day they are constantly fluttering around and are therefore much more difficult to photograph.

Pine Marten crossing a road at dusk

Moonlight

The Moon does not give off light, but reflects that of the Sun. The visible surface of the illuminated part of our satellite depends on its relative position in relation to the Sun and the Earth, which changes constantly. The full moon supplies light at 7 per cent of the brightness of the sun, while at new moon it is no longer visible. Between these two extreme situations the different quarters ensue with decreasing and increasing size and luminosity over the course of the nights. A complete cycle, from one full moon to the next, takes approximately 28 days.

Like the sun, the moon rises and sets. If the full moon rises at dusk and sets at dawn, the first quarter rises in the afternoon and sets before midnight, while the last quarter rises after midnight and sets at the end of the morning. The moon falls behind by about 50 minutes per day in relation to the sun, in other words, 24 hours in 28 days. It is therefore important to check on a calendar which quarter the moon is in, in order to prepare for a nocturnal outing. It is also important to check the weather forecast as a cloudy sky means a black night.

Our eyes adapt to the decrease in light by pupil dilation. On leaving a well-lit building, we can see sufficiently well under a bright moon to make our way through the countryside almost as we can during the day, after a short adaptation phase. In such conditions the use of binoculars is possible, especially models such as 8x40 or, even better, 7x50. Even a telescope can be useful, particularly for watching birds flying high in the sky, when they pass in front of the moon.

Observing the silhouette of flying birds which stand out against the moon is one of the most accessible pleasures for the naturalist who

Lapwings are very active at night

wants to discover nocturnal life. The best viewing conditions are when the moon is low on the horizon, as most birds fly quite close to the ground except during migration. The beginning and end of the night are generally the best times for watching such flights as the birds depart towards feeding zones or return to a daytime roost.

The retina of the human eye is carpeted with two sorts of sensory cells. In the centre, cone cells perceive colours perfectly but need a lot of light to be stimulated. On the periphery, rod cells only allow black and white vision but react to much lower light levels. The low luminosity of moonlight only allows for black and white vision.

Moonlight creates a particular atmosphere. Vision is much more contrasted, lighter patches stand out from darker areas and the details of elements seen against the light are erased. There is no point standing facing a cliff or a bank if it is not lit up by the moon. Ideally the observer would stay in a shadowed zone and watch a lit up zone. Covering up white patches such as the hands and face enhance the chances of a successful watch.

Saturn's rings make it a distinctive image in the night sky.

Starlight

Owls have eyes which perform so efficiently that they only need light from a starry sky to be able to see well. Much less effective, our eyes cannot get by with this low lighting alone, except to spend the night observing the stars with the naked eye or with binoculars. The night sky is a spectacular sight and there are enough constellations to fill an entire book by itself. If you are particularly clumsy and are discovered by all the fauna in the area, you can always make your trip a memorable one by tracking Mars, Jupiter or Mercury. It is possible, to a certain extent, to use natural reflectors which slightly increase starlight. The most efficient is snow, which becomes visible with slightest ray of light. Try it out, you will find that it is much easier to move around in a wood under a starry sky after a snowfall than before. The areas of snow are visible and tree trunks appear in contrast. Without snow, the dark colours of the earth and trunks melt into an indistinct mass.

To a lesser extent, bodies of water can serve as reflectors and increase luminosity, but don't expect miracles. At best, you will be able to distinguish vague silhouettes on the water, and don't count on being able to use your binoculars to try and identify them.

Over the last 10 years, military technology has developed material which intensifies the slightest ray of light to the point of being able to see almost as if it were daylight. Originally, they were developed as armoured car sights, before being miniaturized and made available for the public to buy. There is now a wide choice on the market. First generation night-vision binoculars are limited to amplifying the weak light of a starry sky. They are often heavy, cumbersome and inefficient in terms of battery use. New generation material is more practical and simpler to use, incorporating infrared technology which allows visibility even in pitch black by rendering visible the heat given off by the body. The miniaturization and portability of these monoculars and binoculars has improved beyond recognition and a huge advance was made with the appearance of night-vision head-sets. They have the advantages of being lightweight and comfortably worn on the head, leaving both hands free. To return to natural vision, all that is required is to move them up onto the top of the head.

This equipment enables you to see well up to a distance of 100 to 150 m, with a clear image from 3 m upwards. Their main drawback is that they generate a black and white image, or rather a monochrome one in green and sepia. These technological marvels cost a few hundred pounds and make for easy observations, but they do take a lot of the wonder out of nocturnal outings.

A night out with only the stars to guide you also offers a good chance to develop the use of other senses as alternatives to sight. You may be surprised at how your hearing seems to improve, while odours will seem more significant, more subtle and more varied. This is a good opportunity to practice making a mental picture of your surroundings thanks to the sound of the wind in the trees, the gurgle of a stream, the smell of silt rising from a marsh or the subtle perfume of a flowering honeysuckle.

Torches

The area of light generated by a torch beam proves very useful for recognising the land and exploring the area in close proximity to the observer. It does, however, have the drawback of signalling the presence of the person to many creatures in the area, as well as preventing the eye from getting used to the natural low light of the night. It can be tempting, in order to see properly, to take a halogen bulb lamp, whose white light is stronger than that of the traditional incandescent bulb, which appears yellowish. Such disturbance is forbidden for protected species and is not desirable elsewhere, so a low-power torch is recommended. In certain areas, in the event of frequent nocturnal outings it can be wise to inform the local police so as not to be confused with a poacher.

The impact of the light beam can be reduced by using a red filter. You can easily make one by using a sweet wrapper and an elastic band. Unlike humans, insects see ultraviolet light but not red. With a normal lamp, most insects will vanish into the darkness; with a red light, they are not disturbed as for them it is still dark. You can therefore observe them more easily.

Many nocturnal vertebrates either do not perceive this colour, or see it very poorly. Their high capacity eyes designed for using the

Garden Dormice inhabit houses across continental Europe

slightest ray of light generally do not have cells sensitive to this wavelength. One drawback is that red light has a lower illuminating power than the normal torch beam.

The use of a torch is recommended for observing flora, invertebrates, amphibians and aquatic fauna. Observation of vegetation goes hand in hand with that of insects, as the plants we can consider as nocturnal are those that attract pollinators at night.

Most plants secrete nectar mainly at dusk and the beginning of the night, and/or at the end of the night and during early morning. There are, therefore, times which are not suitable for observing pollinators at work. The fact that temperatures of less than 10-12°C block the insects' activity and that strong winds will prevent them from flying must also be taken into account.

Other nocturnal insects have too varied habits for it to be possible to give precise observation instructions. If you adventure out with no particular aim, scan the ground, mosses, dead leaves, grass, trunks, bark or foliage. Concentrate on detecting movements, as many insects are very difficult to pick out when they are still and flat against a surface. On the ground you may see centipedes and Devil's

Coach-horse beetles running around, dung-beetles flying onto fresh dung or burying-beetles on a corpse. On the grass and foliage, caterpillars, grasshoppers, crickets and beetles are busy nibbling leaves or hunting prey.

If you are looking for a particular species, locate a suitable biotope during the day, and come back at night at the right time to try and observe it. If you want to see a Garden Spider spinning its web, search in high grass. To see stick insects, investigate the tops of bramble bushes. To observe Stag Beetles in flight, find a post on the edge of a wood or a hedge at dusk.

The best time of year for the nocturnal observation of amphibians is in the spring. Breeders are concentrated in wetlands or humid zones where they can lay their eggs, and the males of numerous species give their mating calls so they are much easier to locate and identify. Amphibians can be observed in all seasons except mid-winter, despite the fact that most of them lead a more terrestrial life scattered in various different habitats. Hedges, scrub, woodlands, fields and especially on the banks of wetlands are all good places to search. A nocturnal walk after a storm or rain which marks the end of a long dry spell offers excellent prospects for an encounter.

If you look into a pond or a stream during the day you will not see a great deal. Animals hide and aquatic vegetation and the sun's glare reduce visibility. However, if you return at night and throw white light upon the water you will see, after an initial moment of panic, a swarm of creatures attracted to the light, from insect larva to large carp. Do not try to catch anything, just observe them: fishing with a lamp is strictly forbidden.

Moth traps

It is a well known fact that moths and other insects are attracted to light, but the reason for this phenomenon is not very clear. It would appear that it is due to a disturbance of their tracking and navigational abilities rather than an attraction as such. Entomologists have developed numerous trap models using light to attract and capture insects. Some can be used by the naturalist in order to sample the variety of insect fauna in a given area.

The principle of the trap is simple: on a dark night, or at least in a place with no moonlight or street lighting, a powerful light source is placed in an open area. A piece of white material stretched between two posts, or on a truncated pyramid structure above the lamp, gives off a diffuse light which is much more attractive to insects. They land on the structure and are thus easy to observe and photograph. They avoid getting too close to the lamp and burning their wings. Near buildings, any sufficiently powerful lamp connected to the mains electricity is suitable. Outdoor lighting on a well exposed wall often proves to be a very acceptable light trap for discovering the nocturnal fauna in the garden.

Away from buildings entomologists use special lamps connected to a battery or power unit which give out a lot of ultraviolet light. For those who are just out for the pleasure of discovery, something like a gas camping lamp is sufficient. Easy to transport and use, they have a good track record. Watch out, however, for fire risks, especially in southern Europe or during dry spells. In such cases it is better not to use them. Insects do not fly every night, all the time or in all weathers. The best time is the first half of the night, preferably prior to midnight or about 1 o'clock in the morning. However, it is possible to find

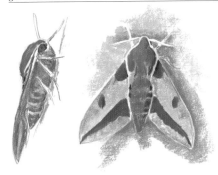

Spurge Hawk-moth
by torchlight

insects from dusk until dawn, as certain species only fly at the end of the night.

The most favourable weather conditions are warm dry nights. Overcast weather, which prevents nocturnal cooling, is particularly favourable, as is stormy weather, as long as the storm hasn't broken. High winds, cold and to a lesser extent rain, are all detrimental to insect movements. The tracking period for insects covers that of flowering plants, from the end of winter with the willows to the middle of autumn with the ivy, but it is especially from the middle of spring to the beginning of autumn that searching is most productive and observations most varied.

The position of the trap must be carefully chosen. Avoid close proximity to well lit places, large scale farming zones and places which are highly exposed to wind and airstreams. The ideal place is both sheltered and at least partially unobstructed, so that the light will carry well. Unless you want to observe end of night fauna, it is advisable to turn off the trap in the middle of the night, allowing the insects to return to their normal business and find safe shelter before daylight. If not they will be exposed to diurnal insectivores.

With thousands of European species, moths are the main event, but other insects which are often hard to see during the day can also visit the trap. With the evening chafers and Stag Beetles come the dung-beetles, burying-beetles, diving and other aquatic beetles. Other aquatic insects come easily to the torchlight, as well as numerous species from wetland habitats, such as mayflies, caddisflies or certain dragonflies, while we must not forget mosquitoes, including of course the unwelcome ones that bite the observer. Among the rarer encounters might be a praying mantis or Mole Cricket.

Finally, installing a moth trap can be a good way to closely observe the hunting techniques of bats that are attracted to the concentrated mass of insects in a small area and which do not appear to be disturbed by the light.

Night lighting
and light pollution

Entomologists who started to reuse moth traps after 1945, following six years of black-outs due to the war, saw exceptional catches on the outskirts of towns. From the beginning of the 1950s the trend was reversed and little was caught in areas where urban lighting was installed. This was the first obvious proof of the damage caused by light pollution. The insects, drawn to the light, were spending their time flying around streetlamps instead of feeding themselves and reproducing. They were dying en masse, exhausted, burnt or taken by predators.

Today, the sodium vapour lamps used are much less attractive to insects than the old mercury vapour ones, which emitted ultraviolet rays. Nevertheless the problem of light pollution has not been solved – artificial lighting is spreading to more areas and the lamps themselves are becoming increasingly powerful.

At night in overcast conditions, even in remote and wild areas, you can locate the closest built-up areas, big and small, by their clear luminous reflection on the cloud ceiling. In densely populated areas pitch black nights do not occur anymore, much to the despair of amateur astronomers. In these areas, even on nights with no moon, it is therefore possible to follow the passage of certain birds whose dark silhouettes stand out from the brown-orange background of the night sky which is drenched in artificial light.

This situation not only poses a problem to insects, but also to other creatures. Many birds regulate their breeding according to the day-night rhythm. Periods of long days and short nights stimulate egg-laying and the rearing of young, as the parents can search for food for longer spells each day. Poultry farmers light up their buildings 16 hours a day for this reason. Some species that are common in urban areas, such as Robin, Blackbird and Starling, show an increased number of annual clutches. This fact, which can appear to be positive, is a cause of imbalance. The demand for invertebrates, which are the main source of food for the fledglings of most of these species, is artificially increased and they become scarcer.

Artificial lighting is also directly detrimental. The number of animal corpses found on the roadsides is proof enough of the damage due to the glare of headlights causing fatal accidents. Continuously lit main roads can also act as insurmountable barriers for species trying to escape the light, which are unable to cross them.

Migratory birds, often flying at night, are sensitive to this riot of artificial light. The agglomerations represented by big spots of light seem to attract them to the point of having actually modified the routes of certain traditional migratory paths.

Many migratory birds spend hours circling lighthouses, sometimes to the point of exhaustion. Others, tired out, become easy prey for various predators. Some die in collisions, although this phenomenon

Brown Long-eared Bat hunting moths around a streetlamp

has largely decreased thanks to simple adjustments such as lighting up the beacon pole or the masts and antenna mounted on it.

The fact that pipistrelle bats have adapted to hunting moths and other flying insects around lampposts is well known. But artificial lighting is not always favourable towards bats living in towns and villages. Suckling females leave their shelter one by one in a continuous flow as night falls. The sudden illumination of street lamps can delay this flight, with the last females leaving very late. As the peak activity for nocturnal flying insects is at dusk, the bats have more difficulty finding food, which leads to a higher mortality rate among their young.

Studies on the impacts of artificial lighting are still very fragmented, although they do show that the negative effects largely outweigh the positive effects on fauna. Notes taken by naturalists merit being noted and retained, to be used at a later date for a wider study on the subject, or for documenting articles which will improve the somewhat meagre bibliography.

Listening

Interpreting the sounds of nature

With nightfall, the decline in human activity brings about a slight decrease in noise pollution; the noises produced by humans and their machines. It can feel like it is nature's turn to speak once more. This is an ideal time for appreciating natural background sounds.

Unfortunately, this quietness is always relative. Roads with heavy traffic produce background noise which decreases at night but never completely stops. Not even two hours can go by without the hum of a car engine, even on the quietest of country lanes. During harvest time, the combine harvester can work 24 hours a day. And if by chance everything is quiet on the ground, an aeroplane flies across the sky.

If we disregard sounds of human origin, the most common noise in nature is the wind in the vegetation. With a bit of practice, it becomes possible to identify the different sounds of the wind as it passes through the leaves of trees in a wood, the needles of conifer plantation or a reedbed on the edge of a pond. It is difficult to describe the differences between them in words, but they are very noticeable when we have the opportunity to hear them.

The presence of water also generates a variety of characteristic noises. The boom of waves breaking on the ocean shore or a large lake is completely different to the lapping of a river, the gurgle of a stream or the din of a waterfall. The sound of light rain can also tell us about

Common Genet

our surroundings. Depending on whether it falls on a canopy of leaves, a pond or on bare ground, the sound differs a great deal.

Listening to nature is one way of detecting the immediate environment in the absence of light and a great deal of information can be drawn from the surrounding noises. The ear is able to distinguish between regular background noise and more unusual sounds. Learning to listen is also learning to concentrate on the right sound – the one which provides the most information. Once you have identified it, the sound of the wind in the rushes will not trouble you again for the rest of the night. On the other hand, unusual, more abrupt noises will provide complementary information about the environment. A dead tree creaking in the wind, the whistling of a gust of air surging into the funnel of a rock face; as many details of the surrounding scenery as the brain can record.

A good proportion of the sounds we hear at night are produced by nocturnal creatures. Calls and songs, the basis of nocturnal animal communication, are dealt with in detail in the following chapters. The absence of animal sounds can constitute information in itself. If calls stop abruptly, something has just happened; the arrival of an intruder has worried the source of the sound.

It might be your fault if your movements are insufficiently discreet and have been detected. Or it could be the appearance of another potential predator. If the crickets at the bottom of the garden have stopped singing, it is probably the neighbour's cat doing the rounds. If the passerines in a hedge suddenly stop singing, it could be a wild predator and the opportunity for a good observation.

If animals can detect your presence by the noises you make when moving, the opposite is also true. To detect a field mouse trotting along the ground requires a good ear, but some animals would make poor naturalists. A Hedgehog rummaging through dead leaves searching for worms and other prey makes an awful racket, which can be heard up to 20 m away. A herd of Wild Boar is even less discreet: in the silence of the night, the trampling, grunting and breaking of branches can be impressive.

On the contrary, like many other nocturnal animals, Roe Deer will only signal their presence once it is too late. Alarmed by potential danger, they will run away not caring about the noise they make. Likewise, the departure of a Coypu will be indicated by a loud splash and that of a bird by the flapping of wings or an alarm call. However, unless you have an extremely sharp ear, you will never hear the discreet genet.

Calls and songs:
animal communication

The most complex songs of birds, amphibians and insects are linked to the breeding season. These are the subject of the next chapter, but many sounds produced by nocturnal wildlife serve to transmit information from one individual to another in everyday circumstances. At night, sound signals are a favoured way for animals to communicate. Far-reaching, with rapid transmission of over 300 m per second (over 1,000 km per hour), they can be easily adapted and thus transmit complex information. As the night air is not subject to heat currents due to solar energy, sounds can be carried further than during the day, although they can be attenuated or deformed by obstacles.

High-pitched sounds are more easily stopped or dispersed in all directions by obstacles than low-pitched sounds. Generally, though, smaller creatures emit more high-pitched sounds. For this reason, the bellow of a stag carries much farther than the contact call of a forest passerine.

All small creatures, whose songs and calls can be stopped, attenuated or deformed by vegetation and obstacles close to the ground, endeavour to sing either in places which are at least partially open. This is why frogs come out onto the banks of a pond, passerines sing from perches in trees or bushes and grasshoppers or crickets often climb up onto the heads of flowers or grasses.

Songs and calls are used to maintain contact between a pair, especially for owls. The young also call, sometimes not very discreetly, to encourage their parents to come and feed them. It is impossible to ignore the presence of a nest of Long-eared Owls; the fledglings' insistent 'squeaky gate' cries ring out at regular intervals and carry far. For individuals defending a territory, another function of song is to indicate presence and make clear to others that the place is taken. This territorial song can be coupled with the parading song designed to attract a female although this is not always the case, especially for owls. Calls also serve to maintain group cohesion for animals living in packs. Wolf packs, with their strict hierarchy, use sound communication to reinforce co-ordination, especially during hunts.

Calls and songs present one major drawback for those emitting them: they are advertising their presence to predators. Any animal with two ears is capable of locating the origin of a complex sound, such as a trill, thanks to the slight difference between the arrival of the sound waves in each ear. To avoid being located it is better to emit the purest and shortest sounds possible, which is often the case for alarm calls.

Long-eared Owl

The Barn Owl, with its facial disc that amplifies sound, is capable of hunting in total darkness using only its hearing skills. Its main prey, small rodents, give themselves away by the noises made when moving, while it occasionally also captures birds and bats.

Some potential prey species, like the Italian Cricket, have solved this problem by becoming ventriloquists. By modifying the position of its elytrons (wing-cases) while churring, it renders the source of the noise difficult to detect. You hear it and the sound seems to come from a bush on your left. It stops then starts again and you think it is in the tree opposite you. It is much easier to locate this insect by using red light than with your ears.

Mating calls

Caution flies out of the window during the breeding season as all sorts of creatures put themselves on display in order to find a partner. Discretion now becomes a defect: what is the point of staying alive if you can't reproduce? Sexual encounters are often initiated by means of noisy outbursts, although often it is only the males that take the risk of attracting predators.

The birdsong season begins in the middle of winter in regions with mild climates, starting with such species as resident Robins, thrushes or Blackbirds. By early spring most remaining species can be heard, and it is not long before they are joined by early returning summer migrants from Africa such as Northern Wheatear and Sedge Warbler. The end of winter sees the start of the amphibian choruses as frogs and toads return to ponds en masse. The main season of bird and amphibian song begins in spring and lasts into early summer. Males sing to conquer and guard their territory, and to attract a female. This continues during nest-building and incubation, but raising young is a full-time job, which is why birdsong disappears little by little as spring progresses. It can start again before the second brood, but with moulting and the beginning of migration, summer is often a silent time. A few songbirds, such as Robin or Blackbird, sometimes sing again in the autumn, but this remains minimal.

In the middle of spring, when birds begin to quieten, the insects come to the fore as the first churring of the crickets is joined over the weeks by that of grasshoppers and cicadas. They will chirp until the heart of autumn, when it finally becomes too cold and they die off. Dusk is the domain of the cold-blooded amphibians and insects. The heat of the day has not yet dissipated and, if conditions are right, individual sounds can be so numerous that they merge together to create an indistinct background noise. Only a few birds, such as thrushes, Blackbird, Robin and Nightingale, infuse the dusk with their song.

The choruses of insects and amphibians gradually disappear during the night. The temperature drop numbs the singers and, like a record player with dwindling batteries, the rapid, intense continuous bursts of song progressively slow down. The notes break up to the point of being perceptible, before stopping completely. It is only on very warm nights that they can last until morning.

The middle of the night is therefore quieter. Nightingales can be heard at any moment during spring and early summer, and owls can be heard almost year-round, although the best time to encounter them is at the end of winter. The Barn Owl's shriek and the Tawny Owl's hoot are perhaps best known, but each species emits a distinctive sound. Many other birds can be heard in the middle of the night, such as the Quail which is a ventriloquist and very hard to locate, and the Grasshopper Warbler with its insistent reeling that sounds similar to its insect namesake. However, the birds' chorus rings out principally at dawn. It starts roughly two hours before sunrise, opening with the likes of Nightingale, Reed Warbler, Skylark and other species that are not afraid of the dark.

Songs are specific to each species, and with a bit of experience they can be reliably identified. Birdsong CDs are an excellent way to learn to recognise the species before going out into the field. Thanks to the miniaturization of CD- and MP3-players, you can even take the tracks with you on site, enabling more certain identification of the songs heard.

It is interesting that species which are externally very similar, whose visual identification is tricky, can be immediately distinguished by

Lynx

their song – Reed Warbler and Marsh Warbler for example. It is a way for insuring reproductive isolation. Ironically, though, confusion can occur between unrelated species. For example, the monotonous flute-like song of the Common Midwife Toad is very close to that of the Scops Owl.

The hidden world of ultrasound

Bats have conquered the night by developing a high performance system for locating prey and obstacles through ultrasound wave reflection. They 'see' with their ears, thanks to a technique which is similar to that used by boat sonar systems.

At best, when young, humans can hear sounds up to a frequency of 18 kHz. Bats emit sounds up to 105 kHz. This is why the flight of bats appears to be noiseless to us, whereas in reality they live in a very noisy world.

Like radios splitting up transistor frequencies, the different species of bats have quite narrow ultrasound emission ranges which vary from one species to the next. They often overlap, but rarely become muddled. These frequencies are, for example, from 10 to 14 kHz for European Free-tailed Bat, which is audible to the human ear, and from 81 to 85 kHz for the Greater Horseshoe Bat. Between these two extremes there are dozens of different species, each with their own specific range.

Other differences between species include the rhythm of the emissions and their reach. Certain bats, such as the Noctule, sweep their environment with powerful calls exceeding a range of 100 m. In contrast, others only explore their immediate environment – Bechstein's Bat's calls only reach 4 m on average. Therefore the call of a bat is, with a few exceptions, a signature enabling it to be accurately identified.

In order to access this precious information, which enables the study and identification of species without disturbing them, scientists have developed ultrasound detectors and several models are now available. The basic principle is that a determined range of ultrasounds enter the appliance and are released in a range which is audible to the human ear. It is possible to adjust the range of frequencies scanned. The price varies from around 20 pounds to several hundred pounds, depending on the model and the technical sophistication.

Like birdsong, identifying bat calls requires practice and a good deal of field experience. The sharpness of the ear and sound memory count too. Bat enthusiasts often record the sounds they hear in order to process them later and make a more certain identification. Again it is possible to find CDs of bat calls which enable you to compare species and put a name to a call with greater certainty.

The world of ultrasound is not confined to bats. Insects produce them as well and a bat-detector is perfectly adaptable for discovering them. One possible use is to compensate for diminished hearing capacity linked to age by rendering their high-pitched stridulations audible – the Speckled Bush-cricket falls into this category. Some entomologists also use the device for locating species with easily audible calls, such as the Alpine Dark Bush-cricket, Long-winged Conehead and Large Conehead, but whose repertoires overrun to a large extent into ultrasound.

The luring technique

Wild creatures call or sing to communicate with each other and exchange information. Over the millennia humans have used this behaviour to attract game or locate it more easily so that it can be hunted. Imitating calls has long been used to lure birds, for example.

Calls to indicate a territory or attract a mate are the basic material for luring. Conversing with a bird is an art. If you use calls and songs recorded on a CD, you must not use the wrong repertoire or the bird will leave or become quiet. You should also give it time to express itself and, as required by human good manners, you should not interrupt it during its own song. Depending on the species, a bird that has been lured could land close to the machine or several metres away. It is up to you to find the best spot to appreciate the performance.

The call you play can attract a bird because it thinks it has found a mate, but most often it thinks that there is a rival that wants to take its place. In such cases it could become stressed and aggressive. A bird which is distressed in this way could potentially abandon its territory and compromise its chances of breeding.

Although acceptable for a scientific study, the luring technique is less justifiable for observations that are solely for pleasure. It should, therefore, remain a rare event so as not to cause damage through overenthusiastic curiosity. It can also be used with certain insects, without the same restrictions noted for

Barn Owls can be attracted by voice imitations

birds. For example, at dusk in front of its burrow the Field Cricket emits a churring to dissuade rival males from approaching. You can incite a reaction and increase the volume and frequency of its churring by simulating the approach of an insistent rival with your machine. Or you can improvise quite easily without using any specialist equipment. The Common Field Grasshopper can be heard from the middle of spring to the middle of autumn during the heat of the day and at dusk. Its stridulation is a characteristic *bzzzt* lasting 0.2 seconds, repeated at the same rhythm in a monotonous phrase which lasts about two seconds.

When two males are close, they launch into a sonorous combat, alternating their stridulations in a vigorous crescendo. It is easy to copy the sound of a Common Field Grasshopper by imitating it with your mouth. Depending on the speed and vigour of the insect's reply, you can appreciate the accuracy of your imitation and hone your technique.

Smell and touch

Nature's odours

With a little practice, odours transported by the night breeze can provide a great deal of information. Calm nights with increased humidity in the air are best for detecting these odours, while windy days render our sense of smell almost useless. Very dry nights, especially during hot periods, offer an array of heady aromatic odours at the expense of the more subtle scents.

Whether pleasant or unpleasant, odours can tell us stories. It is hard to use words to describe a smell, so in this domain personal experience is key. The earth and vegetation emit scents which inform us quite accurately about our surroundings. The smell of damp earth or silt is characteristic of marshes and riverbanks. The proximity of the sea can be detected by the salty air. On the ocean shore, high tide does not provide many odours, but in contrast the smells of silt and seaweed are present at low tide.

On a walk in the countryside, a strong smell of humus signals that the path penetrates into leafy undergrowth. The odour of resin will reveal a pine plantation. Mediterranean scrubland assaults the senses with vegetal odours: at almost every step you will smell something new. Agricultural areas also have their own typical smells and cowpats can be easily distinguished from sheep dung just by using your nose. Less agreeable and less natural is fermented fodder, which has been extensively propagated in farming regions over the last 30 years – its smell carries a long way.

Dung and other isolated droppings, whether from wild or domestic animals, attract dung-beetles from far afield. It is more difficult for humans to detect them using smell unless they are very fresh and near. On the other hand, the unpleasant stench of an animal corpse can be smelt from afar, especially if the weather is hot and the animal is large. An indefinable rotting odour from the undergrowth can signal the presence of a decomposing mushroom. Like dung and carrion, these attract their own particular associated fauna.

Scents originating from the vegetation act as a means of communication between plants and animals. Flower perfume, which is linked to pollination, is a complex subject which is covered in the next chapter. The perfume of fruit is more discreet and less understood. Fruit flesh is offered by the plant as nourishment for a vector creature (often a bird or mammal) in exchange for transporting seeds in their intestines and depositing them far away via their droppings. The plant signals that the fruit is ripe and can be eaten using colour and smell, that is to say that the seeds are mature and ready to be distributed.

The good smell of ripe apples, peaches or strawberries whets our appetite, showing that we understand this plant language. Many wild fruits give off pleasant smells but they are often quite subtle. A nocturnal walk during autumn is an opportunity to discover some of these, and if they are edible why not taste them? However, it is essential to know what the fruit is before putting it in your mouth as some contain dangerous poisons.

Humans leave traces of natural odour under certain conditions. The strong odour of sour urine along a quiet country path could be due to the presence of a car park, meaning that visitors often empty

Moths attracted by the sweet smell of fallen plums

their bladders in the same spot. However, before throwing accusations check that there is no hogweed in bloom in the area – this plant, pollinated by insects attracted by droppings, gives off a smell very similar to that of urinals.

Flower perfume

Plants flower in order to reproduce. Fixed to the ground by their roots, they must use an external agent for transporting the male pollen to the carpel of a flower of another individual of the species. This agent can be water, or more often wind, especially in the case of trees and grasses, but 80 per cent of plant species rely on living vectors to produce these fertilizations, and in Britain and Europe these are almost always insects.

Flowers advertise themselves to insects in a number of ways, the two main ones being colour and odour. Diurnal insects react to colour from afar, and to odours from a short distance. Conversely, nocturnal insects sometimes react to odours from a long distance. Their antenna can act as incredibly sensitive detectors. Thus, the male Oak Eggar moth is capable of locating a female at a distance of three or even five km, solely because of the odour emitted by the latter, which is not perceptible to us. Flower colours play a minor role at night. They serve as a guide at dusk, when the light is still sufficient for insects to see them despite their inferior vision. These twilight flowers do not display dazzling colours or complicated nectar guides, which would be unnecessary. They are often white, pinkish or pale yellow, in short, always a colour which is easily visible in the weak light of dusk. Many nectar-gathering insects are dusk creatures rather than truly nocturnal, like hawk-moths for example.

The most common vector for night flowers to communicate with insects remains odour. Stand under a flowering lime tree on a spring evening. As the light dims, you will smell its perfume begin to intensify.

Convolvulus Hawk-moth
plundering nectar
from geraniums

The perfume can be detected several metres away, particularly if there is light breeze, and this attracts numerous nectar-gathering insects that are flying in the vicinity. Taking a look at the traffic around the flowers, it is obvious that this strategy is very effective. The Sweet Chestnut tree, which flowers roughly at the same period, also smells stronger at dusk, although its sickly sweet odour is slightly more discreet.

Some flowers raise the curtain in the evening before starting to emit their perfume to attract insects. During the day, the Night-flowering Campion presents flowers huddled in its greenish sepals, hardly visible at all. At dusk, the pale pink corolla unfolds and a strong perfume is produced. A closely related species, the Nottingham Catchfly, keeps its white petals rolled up during the day, unfolding them only at nightfall – a process which is synchronised with the emission of its perfume. Probably the most extraordinary display of all is that of the flowers of the evening-primrose, which open visibly at dusk only to wilt the following morning.

Nature at your fingertips

Touch is a sense that is often neglected by humans when exploring the natural world. Unfortunately, nature-watching at night is full of traps in this regard due to the reduced performance of sight. It is never pleasant to have sensitive fingertips stung by a nettle or palms pricked by the sharp prickles of a holly leaf or the thorns of a bramble. So particular caution is required when handling vegetation at night. Don't touch or let someone else touch anything unless you are sure that it is safe.

Sense of touch is especially useful for discovering the world of vegetation with your fingertips. A good starting point is tree bark. No possible confusion could occur between the smooth, large sheets of the cherry tree and the coarse bark of the oak. However, between the two extremes many subtleties exist.

Feeling plants can add to the suite of identification features that can be detected by sight and smell. Of course, it is important to watch out for the dangers indicated at the beginning of this chapter. For example, Hogweed, Giant Hogweed and Wild Parsnip can make the skin touched by their leaves light-sensitive. At the time you don't feel anything, but once exposed to sunlight the skin becomes red and itchy. This reaction depends a lot on the individual and the type of skin. Some are very sensitive, others not.

While we are giving warnings, do not forget that certain wild animals, especially small carnivores and rodents, can transmit some quite nasty viral diseases through simple contact with their droppings. So if you must handle this kind of material for identification purposes use gloves or tweezers, and never handle them with your bare hands.

Talking about touch sensations is equally as difficult as describing odours. The work of botanist François Couplan uses a vocabulary which is rich yet unambiguous. Here it is a short glossary of plant textures which has been borrowed from him:

Clinging: madder, bedstraw and pellitory.
Fleshy: stalks of speedwell, glasswort and butterwort; leaves of purslane and house-leek.
Fuzzy: leaves of Foxglove, Great Mullein and Marsh Mallow.
Glabrous: smooth without hair. Most plants have glabrous leaves.
Hairy: leaves of Hornbeam (underneath), Hazel, Lavender and

Mouse-ear Hawkweed.

Limp: larch and yew needles, leaves of
bryony, Common Mallow and Greater
Celandine.

Prickly: gorse, thistle, scabious, Holly,
Bramble and Blackthorn.

Resinous: the needles of all pine trees,
but also birch leaves and Black Poplar
branches.

Rough: hop leaves.

Rubbery: leaves of water-lily,
Arrowhead, Black Bryony, Lesser
Celandine, Martagon Lily and gentian.

Rugged: leaves of elm, redwood, couch
grass and cow-wheat.

Spongy: bulrush leaves.

Sticky: Alder leaves.

Tough: pine needles, leaves of Holly, Ivy,
Boxwood, bilberry and Mistletoe.

Urticarial: nettle leaves.

Viscous: leaves of henbane.

Moonlit gully

Hides

Daytime preparation

Setting up a hide means being on the look out for an advantageous spot from which to study wildlife. Gaining exceptionally good views of a particular creature or spectacle depends a lot on luck, but good preparation is also important. There is no point in going out every evening hoping to get lucky. It is better to check over the terrain during the day, looking for the tracks and signs that indicate the presence of the animals you wish to observe.

A single, well-prepared hide will give you a far greater chance of success than several nights spent outdoors searching at random, although remember that success can never guaranteed in either case.

Most animals, even if they only come out once night has fallen, leave traces which allow you to pinpoint the places they visit. A fox's footprint, genet droppings, a collection of owl pellets, bark frayed by Roe Deer – the list is long. To reinforce a series of sightings in your memory, it is worth noting down the different clues onto a map to keep track.

Like humans, wild creatures often adopt habits that require the least possible effort. They take the same paths to get around, especially if they are simply going from one place to another rather than hunting or foraging. These paths are often shared by several species, with each finding it more practical to use an existing run rather than open up a new one. People sometimes do the work for them. Rural pathways leading through fields and woods have limited traffic which peaks during daytime. These are often busy throughfares for nocturnal wildlife on their journeys, for example between resting and hunting zones. Numerous signs line the most popular routes, including footprints, droppings and the remains of meals.

In more dense vegetation pathways can be spotted straight away. There is often a main trunk with numerous forks leading off it; a genuine network of paths. Pathways can sometimes follow unexpected routes, for example when a Weasel's trail runs over the top of a wall. It goes without saying that the bigger the animals are that follow the pathway and the more often it is used, the more visible it is, even in bushy vegetation. This is the case for paths used by Red and Roe Deer, Wild Boar, Badgers and many other species.

The regular routes taken by smaller or more discreet animals can easily be seen in areas of low-growing vegetation, particularly in winter, but they quickly close over again when grass or other plants start growing vigorously in spring and summer. For herbivores or omnivores such as Brown Hare, Roe Deer and Wild Boar, traces of feeding sites can often be detected along their pathways. Those of carnivores can be strewn with the remains of meals. These elements are good clues as to the identity of the creature that created the pathway. On riverbanks, Otter, Coypu, Muskrat and Water Vole leave slides – their passageways between water and land – which can be easily spotted.

Bodies of water found in the temperate climes of Britain and Europe do not generally have the same magnetic appeal to wildlife as those in the African savannah, but they do attract wildlife to a certain extent. This is especially true in places that, due to their geology, possess little or no surface water. Such conditions can lead to concentrations of animals at water sources – the ideal situation for a hide.

The canny wildlife-watcher will take advantage of local circumstances when they could result in an interesting experience. For example, human activities during the day can have a direct influence on night-time observations. In spring, hay-cutting is a catastrophe for small rodents. They suddenly lose the shelter of the tall vegetation. The meadows then become a hunting ground for foxes, owls and other predators that take advantage of a food source which is temporarily easier to capture.

A European Nightjar takes a drink while on the wing

How to set up a hide

A hide operates according to a few simple principles: the observer, by remaining still and quiet, and in position for a long period of time, is not detected by the animal and bird community. The reward for making such an effort is that creatures sometimes come very close and give views that will be remembered for a long time.

Patience is the key to success. The amount of time spent waiting is invariably much longer than that spent observing, and there can be no guarantee of seeing anything. There is no point setting up a hide if you cannot keep still or if you start daydreaming after half an hour when nothing has happened. You need to remain permanently alert so as not to miss a passing creature which may only be in view for a few seconds. The ability to stay still and remain attentive depends on the character and capacity of each individual.

A few simple rules should be adhered to when choosing, preparing and constructing a hide. Failure can be down to forgetting the tiniest of details. For a hide to succeed the observer must not be discovered, so concealment is an extremely important factor.

The simplest hide can be created by using a natural feature: a rock, tree trunk or ditch are good places to hide that blend in with the scenery. A piece of ground-sheet slipped into the rucksack will add an element of comfort to a situation which could potentially be very uncomfortable. Such a makeshift hide can be a good solution during an unexpected encounter, in order to make the viewing period last longer without running the risk of being discovered.

If a night outing is planned in advance with a daytime visit to the site, the position of a hide can be determined more precisely. A simple way is to choose an existing structure such as a stone wall, shack, the body of some abandoned farm machinery beside a hedge, or a blockhouse on a sand dune – there could be a number of suitable sites within the territory of the creatures that you wish to study. Easier still is to make use of the numerous hides and observation screens provided in established nature reserves, which are slap bang in the middle of fantastic wildlife habitat.

Of course, such fixed hides do not exist in many of the places frequented by sought-after animals. In such cases a temporary or mobile

Badgers emerge from
the shadows

hide is the answer. The range of models that can be built is extensive and
what is most suitable depends upon the type of observation, the weather
and the materials and resources available.

The car, a mobile hide

The idea of observing wildlife from a car may seem absurd. What
could cause more disturbance and be more noisy in a natural
environment than a motor vehicle? If you take a car into a wild place
with headlights on and engine revving you are certain to disturb the
wildlife. This is not recommended.

Vehicle disturbance is forbidden or strictly limited in the majority of

A Beech Marten at
the roadside

nature reserves and 'off-roading' is discouraged in many other places. Driving along private roads or onto private property should not be done without prior authorisation. On the Continent, poaching is often carried out using projector lights shone from four-wheel drive vehicles. In the last few years, fairly restrictive regulations have been passed prohibiting people from searching for or following game by car, especially off road with projector lights. In such situations it can be difficult for the wildlife-watcher to convince the police of their good faith, particularly when the latter find themselves regularly confronted with poachers.

The car can, however, prove to be a high performance tool for observing wild fauna with almost no disturbance. Britain and much of Europe have been densely populated by humans for a long time. The network of roads and paths marked out across the countryside is so dense that the wildlife populating it is used to traffic. Even though road-kill deaths remain extremely common, many wild animals do not consider the car to be a threat. Light from headlamps, the noise of the engines and speeding vehicles have always been part of their daily lives. In contrast to a human silhouette, which is immediately associated with danger, the car is not seen as a dangerous object. Cars follow the same paths night and day, never chasing animals (except for poachers). Even when off the road, a vehicle is rarely alarming. Farmers often work at night, particularly during harvest time. Deer or Wild Boar disturbed by the arrival of a tractor sometimes move only a few metres away and then continue going about their business as normal. In the same way as farm machinery, the car is not considered to be a threat. Chance wildlife encounters are a common occurrence while driving at night, whether it is a fox moving with stealth, a hunting owl or a toad crossing the road. These can rarely be transformed into full observations because we may be in a hurry, it may be unsafe to stop or we may lack the correct equipment to make the most of the situation.

Why not make a car journey specifically with the intention of locating the spots frequented by wildlife? The outing can be random or organised in a circuit, taking in a range of habitats and interesting sites in the area. One of the aims is to have chance encounters in unexpected places. However, if long distances are covered and a vast area is explored in one single night, the probability of an encounter between a car randomly driving and an animal roaming its territory remains low.

The car can also be used as a classic hide. Parked on an embankment, it protects the observer from the cold, wind and rain, offering a relatively comfortable seat from which to watch, for example, the birds on a lake. Likewise, parked beside a field or meadow, it may be possible to watch a fox pouncing on a rodent or a group of Wild Boar searching for food without disturbing them. Of course you must remain in the car – the appearance of a human silhouette is a danger signal that will trigger panic.

Finally, you can use the car to drive to a wildlife site during the night and sleep. This gives the animals a chance to forget about you and you can discreetly get out of the car just before dawn to begin investigating the surrounding area. This method is less liable to cause disturbance than arriving on site at the last moment, and the night is spent more comfortably than if sleeping in a rough temporary hide.

Attracting animals

Since the Palaeolithic era hunters and trappers have used bait to increase their chances of success. Naturalists can employ this efficient technique to attract a variety of animals to one particular place and improve their chance of some interesting sightings. In effect, you can lure the animals to your hide rather than taking the hide to the animals. To do this you must make the area outside your hide attractive to wild animals.

Place one or two whole dead chickens 20 to 30 m from the hide, being careful to secure them properly so that foxes cannot run off with them. A length of wire or a strong cord attached to the base of a bush or tree trunk is generally sufficient and will ensure that your bait will last for several nights. In a very open area, take a mallet and stake to knock into the ground. You can avoid fixing corpses by using meat for dogs bought in a supermarket and cut up into small pieces. The mass of rotting meat gives off sufficient odour to attract scavengers such as fox, Badger, Hedgehog and Wild Boar. It may bring in a few mustelids, but they will only be able to eat on site. The dog meat option is very practical in winter when the cold prevents decomposition. If you cook it in oil, the smell will carry a very long way.

You can also use eggs, fish or tinned dogfood. Bait must be laid out undercover; if not, Magpies and other corvids are likely to spot and plunder it. Hazards of this method on the Continent (or in a few parts of southern England these days) include herds of boar. If a group of these gluttons appears you will have to start all over again as they are opportunist scavengers with huge appetites and will clear an area in minutes.

A Brown Bear carefully approaches bait in Slovenia

You can also use the baiting technique for insects. Necrophagous creatures, which live on corpses, and coprophagous creatures, which live in dung, locate these food sources by odour and fly in. To protect themselves against predators their movements take place principally at night. It is best to use very fresh dung or droppings as bait. The bigger the pile, the more attractive it will be. To attract necrophages, take a fresh corpse; the species that will approach will differ depending on the state of decomposition. To avoid having your corpse-bait stolen by a scavenger, put it in a small basin on a layer of sand or earth and hang it on a branch or on three large crossed stakes at a height of 1.5 m above the ground.

It is best to set up these baits on quiet, warm, overcast midsummer nights, which is when insect movement is at its greatest. Avoid cold or windy nights or continual rain. You can check for insect arrivals with a lamp, preferably one with a red light filter so as not to disturb them.

Preparation of a sweet liquid is an ideal way to lure moths. Raw or cooked fruits make a good base: overripe bananas or peaches or a jar of apple puree are effective. Add a large quantity of sugar, such as honey, syrup or granulated sugar, and a small amount of alcohol (beer, wine or rum) to reinforce the aroma and therefore its appeal to the insects. Stir the mixture well and mash up the fruit pulp to obtain a liquid with the consistency of pancake batter. A little water can be added if it is too thick. If there is time to leave the mixture to settle and begin fermenting, so much the better.

Shortly before dusk, brush the trunk of a tree with this mixture at about eye level. Now all you need do is sit back and watch the show. The best weather conditions for the approach of moths are the same as those recommended above for necrophages and coprophages. The sugary solution will also attract different species of insects during the day.

Walking

Tracking a particular animal

Instead of waiting in one spot for an animal to appear, you can decide to go actively looking for it. The technique is preferred by some naturalists because it replaces what can be a long, boring, joint-stiffening period of waiting with a more active and physical experience. However, making a successful observation is generally reserved for those who are adept at the silent approach.

The method is a good way to observe Hedgehogs, even for beginners. Well protected by its spiny coat, this is one of the few European mammals which does not react to danger by fleeing. Instead it comes to a halt and rolls into a ball, making it very difficult for a predator to gain a hold. This defence mechanism is very effective – only the likes of a Badger or Eagle Owl can force it to open. Therefore Hedgehogs tend to be very confident and roam the countryside indiscreetly. They only react to human presence when in very close proximity, and always by first coming to a standstill. A night-time trip to track Hedgehogs can often be fruitful and can be carried out in groups, even with quite large numbers of people. However, success can never be guaranteed and it is important to know a little about the animal's habits in order to increase your chances of crossing its path.

The Hedgehog's daily routine is characterised by two busy periods

A Hedgehog captures a toad

of activity and hunting, at the beginning and the end of the night with an intervening calmer period in the middle of the night. There is no point in looking for it between November and March as it hibernates and will be comfortably settled in a bed of grass and dead leaves. The species frequents meadows, fallow farmland, field borders, and glades. It is often found in gardens on the outskirts of towns and in coastal sand dunes, but does not like pinewoods or wetlands. Farmland with plenty of hedgerows is its favourite haunt.

Other nocturnal creatures are noisy enough to be located by ear at quite a distance, although none are quite so confiding as the Hedgehog. Owls often communicate with each other through loud calls and hoots, which enable the observer to get closer and increase the chances of a sighting. Among the mammals, Wild Boar are noisy and not particularly shy. A group pushing its way through the undergrowth or ploughing up a meadow in search of roots can be heard a long way off. This can be quite an impressive experience in total darkness, although it can upset certain people, especially children. Unlike Hedgehogs, if you try and get too close to boar they will run away. With a good knowledge of the terrain and species, you can anticipate the movements of your quarry and position yourself in a place they will visit of their own accord. It is a way to more or less guarantee of a good observation, sometimes quite close up.

This technique of concealing yourself in a place regularly visited by a particular animal has been called the 'reverse approach'. It is the animal which approaches you, and not the reverse. It is a technique for experienced observers only and can require a range of skills and a good deal of knowledge depending on the species. Robert Hainard, for example, used to hang up a hammock in the woods close to a boars' den. The boars forgot about him and at dawn he found himself with a front-row ticket as the animals returned to their shelter. The 'reverse approach' can also work with birds, particularly waders, which sometimes come to within a few metres of the observer on a rising tide.

Unless you use the moth traps or food baits such as those discussed above, it is a waste of time to wait at a fixed point for insects to come to you. Some species only range over a very small area so it is necessary to actively search for them.

Walking is the best way to locate nocturnal insects which give off light, for example Glow-worms and Fireflies, or produce sound, such as crickets and grasshoppers. In the case of the last two, the use of an ultrasonic detector for locating where the sounds come from is an undeniable advantage.

Taking advantage of romance

The self-preservation instinct, perhaps coupled with negative encounters, renders most nocturnal animals extremely suspicious of humans. However, they often lower their guard during the mating season. The aim of their existence is to reproduce and they need luck on their side to meet a partner. Songs, calls, displays and bright colours; anything goes in order to try and appear more attractive than the neighbour.

It is the males in particular who risk exposing themselves to potential predators at this time, from crickets and toads through to deer and birds. The well-prepared naturalist will know exactly the best times on the calendar for going out in search of particular species, vastly increasing the chances of making some memorable sightings during a nocturnal outing.

The end of winter, for example, is a good time for listening to the songs of owls. It is possible to hear them throughout the year, but activity is particularly frequent during this period as claims are being staked on territories prior to the breeding season. During this same period many species of amphibians begin to move from their hibernation sites to their breeding areas. Male frogs and toads soon begin to sing beside wetlands. If the weather conditions are right, you are almost certain to encounter them.

Spring sees the large scale hatching of mayflies along riverbanks. Huge numbers of these insects hatch almost simultaneously, and during the short time that they live they saturate the area with their numbers and decrease the risk for each individual of finishing in the stomach of a predator. These hatchings occur at the beginning of the night on calm, mild evenings. In the past it was a spectacular sight,

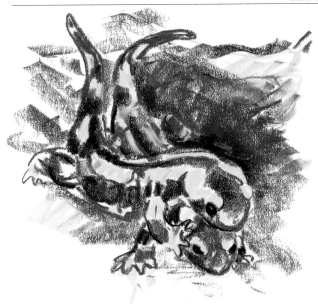

Fire Salamanders fight over a mate

with layers of bodies several centimetres thick covering the ground the following day; today their numbers are much more modest, with river pollution and development perhaps responsible for reducing the number of larvae. Remaining on the rivers, in late spring you can search for spawning Allis Shad.

Among the mammals, the roar of the Red Deer is one of the most impressive demonstrations of the instinct to reproduce. At the beginning of autumn, stags that normally live in groups isolate themselves in a part of the forest. They roar at nightfall, both to attract females and mate with them, and to intimidate their rivals. As there is not enough space for all the stags, roaming individuals regularly try to chase off those who are holding territory. The stag on the point of roaring stretches out its neck, flares its nostrils, then utters its long guttural cry which carries far into the forest. This very deep sound can effectively pass through tree trunks without being stopped or distorted. The roar of a deer not quickly forgotten and many people hearing it for the first time are surprised by its power. Although Red Deer are easier to locate during the rut, they remain very shy and only a quiet and discreet approach will allow you to get within close range.

Inspecting the vegetation

Walking along a path a night, with the help of a torch you can observe many creatures which come to feed or hide in the vegetation. This is the only truly effective method for discovering a range of species with secretive habits which live at night in order to escape the attentions of predators. The beauty of nature, however, is that the existence of a food source brings about the appearance of species ready to exploit it, and you will also come across nocturnal predators that are more difficult to find during the day.

Hiding during the day and coming out at night is a strategy widely used by caterpillars. These larvae of butterflies and moths, which spend much of their time eating, have few defence mechanisms apart from poison or camouflage. As nocturnal insectivorous birds and bats chiefly attack flying insects they are relatively safe at night, although other predators can hunt them, especially rove-beetles.

If you are looking for a particular species, locate its host plant during the day and look in particular for those which have had their leaves eaten. Then return at night to look for the culprits on the stalks and leaves. If you just want to observe nocturnal caterpillars in general, inspect the foliage around the base of trees and bushes which are the favoured food-plants of a wide range of species: oak, willow, poplar, wild rose and bramble, for example. Certain herbaceous plants such as dandelion, grasses, docks and clover are also good in this respect. In the garden, most vegetables are prone to nocturnal caterpillars, especially cabbage, chard and sorrel.

Insects disappear in bad weather, although Pine Processionary caterpillars are a remarkable exception. The larvae of this moth, which is gradually increasing its range northwards every year, shelter in a collective silk nest during the day. These pouches can be easily spotted on many ornamental pine trees, notably the Austrian Pine.

If you find a pouch quite low down, return at nightfall when the weather is calm and mild at the end of autumn or during winter. With the help of a torch you will be able to see the caterpillars coming out of the nest in single file, spreading out across the surrounding branches to devour the pine needles.

Many other insects can be tracked down using the light of a torch. In continental Europe you can hunt for stick insects in the tops of bramble bushes. In southern Europe, the Praying Mantis shelters in grassland with scattered low bushes during the day and hunts at night. A large number of grasshoppers also have nocturnal habits, with some like the Oak Bush-cricket gnawing leaves and others like the Great Green Bush-cricket devouring other insects.

A wide range of invertebrates comes out night. Snails, slugs and centipedes animate the leaf litter and vegetation with their activity, while a torch with a red filter may reveal a spider building or rebuilding its web.

Don't hesitate to search inside dandelion heads, the folded petals of a buttercup, the tightly coiled corolla of a bellflower or the bell-shaped flowers of heather. These provide refuges with a temperate microclimate which is favoured by flies, bees, wasps and small beetles.

Wandering at will

Wandering at will means looking at the world around us just as a poet would. Poetry is all around us in nature, cropping up everywhere even though we are not always ready to perceive it.

When trying to identify a song, follow a shadow or simply look at our surroundings and try to understand what is around us, our brain and our senses are occupied. They are homing in on a single thing, meaning that the rest of our immediate environment is not analysed or perceived.

The magic of the night is not just about the creatures populating it. Constant searching for a sighting of a particular species focuses the mind on one goal. Wandering at will, with no preconceived idea of what the night will offer, opens the mind to all types of sensations, and consequently to all types of observations.

By surrendering yourself to what comes spontaneously, you will be more able to feel nature's pulse and to see how life expresses itself in thousands of different ways. You will be more in tune with the atmosphere, which can be so different on a summer's evening in the mountains, an autumn night in the forest or a spring dawn on the banks of a lake. You will be able to taste the poetry of the shadows, sounds and smells. You can wonder at the beauty of a night sky and discover constellations and shooting stars.

A nocturnal walk simply for exploration allows us to do things that would not be possible during a trip designed specifically for finding a particular type of creature. For example, a bicycle ride in the moonlight on quiet country roads or tracks permits a much larger area to be covered than is possible on foot.

If watching at dusk has the advantage of being a continuation of daytime activities, enjoying the sight of the dawn can be psychologically more motivating. Seeing dawn during the summer means getting up unusually early but it is invariably worth it. At dawn or dusk the arrival or departure of the sun is always accompanied by numerous wildlife spectacles, big and small, that will increase the pleasure of the outing. Having an open mind during a night outing sometimes means letting the animal which is inside us express itself. New sensations and new feelings invade us and anxiety can overtake us at the sound of an unusual or very close noise.

A sudden release of adrenalin which urges us to flee is an instinct which comes from deep down – it is a survival reflex and a defence mechanism against unforeseen circumstances and is something which rarely comes into play in our lives today. Immersing ourselves in the night is therefore a way of rediscovering instincts that have been suppressed by modern life.

Fox in the moonlight

II

IDENTIFICATION

GUIDE

Flowers

It may seem odd to refer to nocturnal plant species. Plants are present in the same spot night and day and, as the vast majority use photosynthesis, they could easily live in eternal daylight whereas a never-ending night would kill them.

The nocturnal adaptation of plants is only indirect. Some species have evolved to form partnerships with nocturnal animals, or more precisely nocturnal insects. This comes to the fore during an important phase in their lives: reproduction.

As plants cannot move, they need to use an external vector so that the male element can fertilize the female. Some species use water or the wind and others use insects. As many insects are nocturnal, certain plants have adapted their flowers so that they open during periods of low or non-existent light. They tend to use light, solid colours and heady perfumes to attract their guests.

The flowers covered here are just a small selection to illustrate the different degrees of adaptation to the night. Many others can be discovered in the field.

No native European tree or shrub is totally specialized in terms of pollinating its flowers via nocturnal insects. Some, such as oak or hazelnut, rely solely on the wind. Many, such as lime and chestnut, have retained a high pollen emission which permits partial pollination by the wind despite the fact that they produce nectar to attract insects. However, a look at flowering trees and bushes at nightfall is always worthwhile. Here are four of the most interesting species.

Goat Willow *Salix caprea*

Salicaceae
Height: up to 12 m
Distribution: all of Britain and Europe.
Flowers: late February to April.

Like other members of its family, this willow opens its catkins early in the season before the leaves appear. Although not specially adapted to the night, the nectar is easily accessible and, being a rare resource at the end of winter, ensures the trees are frequented by numerous insects at dusk and during the night. Certain species of moths which hatch early in the season can be seen most easily on these trees.

Wild Privet *Ligustrum vulgaris*

Oleaceae
Height: up to 3 m
Distribution: all of Britain and Europe.
Flowers: May and June.

The privet's clusters of shallow white flowers attract numerous insects during the day. Their white colouring makes them highly visible in the declining light of dusk, while their sweet smell renders them attractive to certain species of moths which are regular visitors.

Sweet Chestnut *Castanea sativa*

Fagaceae
Height: up to 30 m
Distribution: native to southern Europe,
introduced elsewhere including Britain.
Flowers: June and July.

Originally from the Mediterranean, this tree has been cultivated since ancient times for its edible fruit. Widely introduced further north, it is particularly partial to acidic soil. Its southern origins are reflected in that it has retained an aversion to the cold, coming into flower very late in the season at the end of spring or in early summer. The nectar is easily accessible and its sweet, slightly sickly odour attracts many insects during the day and at night. Stag Beetles are often drawn to chestnut trees, even though they do not gather nectar from them.

Small-leafed Lime *Tilia cordata*

Tiliaceae
Height: up to 30 m
Distribution: most of Europe, including England.
Not in high mountains.
Flowers: June and July.

The lime, or linden, with its shallow flowers and easily accessible nectar, is visited during the day by numerous insects, including beetles, bees, wasps and butterflies. Some of these remain on the flowers at night, but at dusk the heady odour given off by the flowers is reinforced in order to attract moths.

Climbing plants are adapted to attach themselves to other vegetation rather than producing their own support structures. A few of the most common species are very attractive to nocturnal insects. Some have specialized flowers to attract them, which are very light in colour, give off a strong smell at night and are often funnel-shaped in order to reserve access to nectar for the long proboscises of moths. Some species are pollinated during the day as well as at night.

Hedge Bindweed *Calystegia sepium*

Convolvulaceae
Height: up to 3 m
Distribution: all of Britain and Europe.
Flowers: June to October.

Clinging to tall herbaceous vegetation, bushes or fences, bindweed is very common in gardens and on farmland. Its flowers open in the morning and are visited by numerous insects. Some close before nightfall, while others remain open and are well adapted to nocturnal pollination thanks to their pure white colouring which is visible even in very low light.

Honeysuckle *Lonicera periclymenum*

Caprifoliaceae
Height: up to 6 m
Distribution: much of Europe, including Britain.
Flowers: June to September.

Several species of honeysuckle exist in Europe, including this species which is common in woodland and also in gardens, having produced several horticultural varieties. The flowers are pale in colour, powerfully scented with a long tube and are adapted to dusk and night-time pollination by moths, particularly hawk-moths.

Traveller's Joy *Clematis vitalba*

Renonculaceae
Height: up to 30 m
Distribution: throughout almost all of Europe.
Flowers: June to August.

This vigorous clematis is common in hedges and borders and can become very invasive, especially if it benefits from agricultural fertilizer. Its shallow, white flowers with their bunched stamens attract insects during the day as well at night, but their sickly scent becomes stronger as night falls. Noctuid moths such as Angle Shades and Silver-Y come to gather nectar at dusk, and earwigs graze on the stamens during the night.

Ivy *Hedera helix*

Hederaceae
Height: up to 30 m
Distribution: almost all of Britain and Europe.
Flowers: September to November.

Ivy is the most common creeper in Britain and Europe. The simple, shallow flowers are clustered in rounded umbels and open in autumn. They provide abundant nectar which is easily accessible to most insects and is popular with flies, bees, wasps and butterflies. The plant is not particularly adapted to night-time pollination, but moths with few other nectar sources during this season flock to it at dusk.

The campion family encompasses several species which are widespread in Europe. These have evolved in such a way that their features enable the curious observer to guess the time of day that pollination takes place. This can be done by studying the form of the flowers, their depth, colour and perfume. This section covers only a few crepuscular species because many others are essentially pollinated by diurnal insects such as bees.

Red Campion *Silene dioica*

Caryophyllaceae
Height: up to 1 m
Distribution: most of lowland Britain and Europe.
Flowers: March to September.

With its relatively shallow flowers which open during the day and its bright pink-red colouring, this species of campion is perfectly adapted to pollination by diurnal insects, especially flies and bees. However, it is also frequented by moths at night – perhaps the first signs of nocturnal adaptation.

Bladder Campion *Silene vulgaris*

Caryophyllaceae
Height: up to 60 cm
Distribution: throughout Britain and Europe.
Flowers: March to September.

The pure white flowers stay visible even in the low light towards the end of dusk, making them easier to spot in the gloom. They are popular with crepuscular moths, notably the Silver-Y. White Campion, a closely related larger species, has taken a further step towards becoming a nocturnal specialist with its strongly scented flowers that open in the evening.

Nottingham Catchfly
Silene nutans

Caryophyllaceae
Height: up to 60 cm
Distribution: most of Europe.
Flowers: June to August.

This species is quite common in dry areas and has deeply divided whitish petals with lobes which remain coiled during the day revealing a greenish underside. At this time the flowers are barely visible to insects, although bees do visit and pollinate them. At dusk, the petals uncoil and the flower unveils a wide, white corolla which is very visible. It also emits a powerful perfume that attracts moths, particularly noctuids.

Night-flowering Campion
Silene noctiflora

Caryophyllaceae
Height: up to 40 cm
Distribution: all of Britain and Europe.
Flowers: June to September.

This campion prefers dry, sandy soils and is self-propagating and widely naturalised throughout Europe. The flowers are a delicate pink colour on the upper side of the petals, and a yellowish green on the underside. The flowers close during the day, when they become difficult for pollinators to see as they merge in with the colour of the leaves. They only open at dusk, becoming easy to spot thanks to their pale colouration and powerful perfume which attracts moths.

Greater Butterfly Orchid
Platanthera chlorantha

Orchidaceae
Height: up to 80 cm
Distribution: most of Europe.
Flowers: May to August.

This orchid is widespread but never common. Its aromatic flowers are narrow and deep. This is an adaption to pollination by night-flying moths, which can be seen with pollen stuck to their heads, often on their eyes.

A closely related species, the Lesser Butterfly Orchid, attracts the same insects, but the pollen sticks to the proboscis. This substantially reduces the risk of cross-pollination. Other species that are interesting to observe at dusk are the Pyramidal Orchid and the Fragrant Orchid.

Soapwort *Saponaria officinalis*

Caryophyllaceae
Height: up to 90 cm
Distribution: most of Europe, introduced to Britain.
Flowers: June to September.

A close relative of the campions, this species prefers rich, damp environments. Originating from central and southern Europe, it has been widely cultivated as a substitute for soap due to its high saponin content. A double-flowered horticultural form has lost a large part of its attraction for pollinating insects.

The large, deep, pale pink flowers give off a sweet fragrance at dusk. They are frequented by moths, which can reach the nectar thanks to their long proboscises and at the same time ensure pollination.

Dame's-violet *Hesperis matronalis*

Brassicaceae
Height: up to 1 m
Distribution: throughout almost all of Europe.
Flowers: April to July.

This species prefers rather humid climates, where it can be found in meadows, hedgerows and field borders. It is quite rare in the wild. Dame's-violet probably originates from southern Europe, but has become widespread throughout the continent through cultivation in gardens since the Middle Ages, and it can often be found in a semi-wild state close to towns and villages.

Its flowers are white to pale violet, occasionally yellowish, and attract numerous insects during the day, especially butterflies. At nightfall, its subtle perfume becomes more intense in order to attract moths.

Borage *Borago officinalis*

Boraginaceae
Height: up to 40 cm
Distribution: native to south-east Europe, introduced elsewhere.
Flowers: March to October.

This plant, which is used for food and medicine, originated in the eastern Mediterranean basin and spread westwards with the Moors who brought it to Spain. Widely cultivated and naturalised, it can be found in gardens as well as wild places such as hedgerows.

The drooping flowers only offer nectar to insects capable of hanging upside down, such as honey bees and bumblebees. This species is not at all adapted to the night, but as it flowers relatively early in the season when resources are scarce it is visited by the Silver-Y moth, which gathers nectar without alighting.

Gardens are a great place to observe nocturnal or crepuscular nectar-gathering insects. They harbour exotic plants whose nectar resources (but not leaves) can be exploited by the native fauna. A flowerbed laid out near the house will make life even easier, but be careful to choose varieties with simple flowers as these are the only ones to produce a lot of nectar.

Four O'clock *Mirabilis jalapa*

Nyctaginaceae
Height: up to 80 cm
Distribution: exclusively in gardens.
Flowers: July to September.

Originating from warm climates of the Americas, the Four O'clock does not begin growing until May as it is averse to the cold. During summer it will grow vigorously and soon flower if it has sufficient water. The large, flared funnel-shaped flowers only open in the evening, and give off a delicate, pleasant perfume. Selection by horticulturalists means that today there are many different coloured varieties of flowers, including white, yellow, pink, red and variegated. All attract moths, especially hawk-moths, which come in abundance to visit them as soon as they open.

Thorn-apple *Datura stramonium*

Solanaceae
Height: up to 1 m
Distribution: central and southern Europe.
Flowers: July to October.

Originating from the Americas, this was introduced as an ornamental plant but escaped from gardens to sporadically colonise wild habitats such as waste ground and fields, especially where the soil is rich. It is very poisonous.

The large, drooping, horn-shaped flowers can reach up to 8 cm in diameter and are usually white, but can sometimes be pale violet. They are highly aromatic and attract moths, notably hawk-moths.

Common Evening-primrose
Oenothera biennis

Onagraceae
Height: up to 1.2 m
Distribution: introduced into almost all of Europe.
Flowers: May to August.

This is one of several difficult to distinguish, closely-related species that originated in North America and were introduced to gardens in Britain and Europe. Some colonised wild areas, in certain places becoming invasive.

Evening-primrose flowers are pale yellow and sweet-smelling – they open at nightfall only to wilt the following morning. The opening of the flowers is rapid enough to be seen with the naked eye.

Phlox *Phlox* sp.

Polemoniaceae
Height: up to 60 cm
Distribution: exclusively in gardens.
Flowers: April to September.

These hardy plants are a common sight in flowerbeds and borders. Most phlox that are grown in our gardens originate from North America and are hybrid varieties selected for the beauty of their flowers and because they grow easily. Colours range from white to violet and include all the subtle nuances of pinks and reds. As dwarf varieties flower earlier, it is possible to stagger blooms over a long period of time.

Phlox is particularly popular with moths at dusk and during the night, and hawk-moths are often regular visitors.

Tobacco *Nicotiana tabacum*

Solanaceae
Height: up to 2 m
Distribution: exclusively in gardens.
Flowers: July to October.

Originating from South America, tobacco requires warm temperatures in order to grow and it can only be sown late in the season. Several species, including this one, are cultivated ornamentally, while others are grown to produce smoking tobacco. The deep, narrow flowers vary from white to purple depending on the species and variety and can only be visited by insects with a long proboscis. The flowers close up in hot sun, but open most extensively in the evening when they give off a perfume that attracts moths, especially hawk-moths.

Petunia *Petunia* sp.

Solanaceae
Height: up to 60 cm
Distribution: exclusively in gardens.
Flowers: July to October.

A favourite among gardeners, this South American plant has been highly selected and crossbred for the last two centuries. Today, the varieties available on the market are essentially hybrids. Their cultivation is easy in all types of soil, even dry, as long as there is enough sunlight.

The medium to large flowers have deep funnel-shaped petals and their colours vary from pure white to purple, passing through yellows, pinks and violets. The petunia is another plant that is favoured by hawk-moths, which are able to take its nectar with their long proboscises. Certain varieties have a sweet perfume.

Lilac *Syringa vulgaris*

Oleaceae
Height: up to 5 m
Distribution: central and southern Europe.
Flowers: April and May.

Originating from China, the lilac arrived in Europe via Turkey. It is often planted in gardens and can frequently be found naturalized along railway embankments and on wasteland. The colour and form of the flowers has been greatly adapted by horticulturists in many of the varieties, but it is the original old strain with single flowers which harbours the most nectar. With its white or pale violet flowers, strong sweet scent and nectar available via a narrow tube, the lilac is often visited by butterflies during the day and moths at night.

Buddleia *Buddleia davidii*

Loganiaceae
Height: up to 5 m
Distribution: introduced into Britain and much of Europe.
Flowers: June to August.

Originating from China, this bush was only introduced into Europe at the end of the 19th century. It soon escaped from gardens, spreading to wild areas where it can sometimes become very invasive. It is a pioneer species which prospers in wastelands. Its white to violet flowers, ending in a narrow tube, limit the number of species which can gather nectar to those with a long enough tongue or proboscis, so butterflies and moths in particular are attracted to it. It gives off a rather strange smell of fruit and is as attractive at dusk as during the day, particularly the white varieties.

Invertebrates

More than 90 per cent of all living creatures are invertebrates, and nearly three-quarters of them are insects. With such a large number it is impossible to draw up an exhaustive list of nocturnal invertebrates. Even if we disregard minuscule species, it is still an enormous task. For example, there are close to 4,500 species of moths in Western Europe alone.

As a consequence, the following pages primarily cover the larger or more striking species whose crepuscular or nocturnal habits are better known and more easy to observe. Emphasis has been put on two orders of insects in particular; grasshoppers and crickets (orthoptera), which often stridulate at night, and moths, the majority of which are specifically adapted to nocturnal life.

Recognising invertebrates is best done by sight, using a lamp with red light if possible, as they do not perceive this colour. Orthoptera can be identified by ear. The use of light traps or sugary solutions painted onto a tree or a post can encourage moths from the surrounding area to come to the observer.

Jersey Tiger moth

MOLLUSCS

Land molluscs have retained the characteristic moist skin from their aquatic ancestors. They move by generating a layer of mucus, which results in considerable water loss. This explains why they are generally only active at night or in wet weather. A nocturnal existence also enables them to avoid certain predators. Hundreds of species live in Europe, inhabiting all areas which are sufficiently humid. Two of the most common species are given as examples.

Black Slug *Arion ater*

Gastropod
Length: up to 15 cm
Distribution: central and western Europe, including Britain.
Observation: March to November.

This species is common in a large variety of habitats, including gardens, woods, farmland and grassland. It varies in colour from brick red to black, passing through shades of orange and grey. Some experts consider the red morphs to be a different species, the Red Slug (*Arion rufus*). It can be found on the ground or on low-lying vegetation.

Garden Snail *Helix aspersa*

Gastropod
Length: shell up to 4 cm in diameter
Distribution: Britain and much of Europe.
Observation: March to November.

on low-lying vegetation in gardens and at roadsides. In winter and high summer it shuts itself away in its shell, taking refuge under a stone, in a hole in a wall or in another similar shelter.

Despite the fact that it frequents a variety of habitats, including woodland and dunes, as its name suggests the Garden Snail is probably the member of its family that lives closest to humans. It is common

ARACHNIDS

Black Scorpion *Euscorpius flavicaudis*

Scorpiones
Length: 2.5 to 3.5 cm
Distribution: mainly southern Europe. A few isolated colonies in Britain.
Observation: May to November.

This small scorpion can be recognised by its gleaming black body that contrasts with its yellow legs and tip to the tail. It frequents scrubland and other stony places, including human dwellings. Its sting is not dangerous. A few other species can be found in Europe, particularly in wilder areas.

Scorpions are nocturnal and they emerge from their daytime hiding places to hunt spiders, centipedes, insects and sometimes even other scorpions, which they seize between their pincers. They are at home on the ground or on walls.

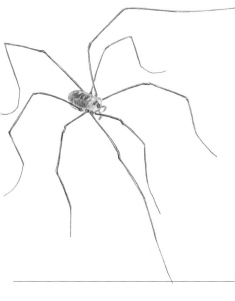

Harvestman *Leiobunum rotundum*

Opilion
Length: 4 to 6 mm
Distribution: all of Britain and Europe.
Observation: July to November.

There are about 30 species of harvestman living in Europe and this is one of the most common. They are found in habitats including gardens, forests, meadows and hedges.

The harvestman can be seen during the day, but has primarily nocturnal habits. They can be found in low-growing vegetation and tree foliage, and the smallest species frequent leaf litter and moss. They have a varied diet and, as well as taking small live prey, they are partial to sugar and can be attracted by nectar traps intended for moths.

Stone spider *Drassodes lapidosus*

Arachnida
Length: 1 to 2 cm
Distribution: throughout almost all of Europe.
Observation: March to October.

This species, the biggest and most common of its family, frequents dry, open places where it shelters under stones or other objects. The body is brownish and the abdomen covered with fine, velvety hairs. As it reaches maturity early in the season, it can be seen from early spring.

Stone spiders take refuge in a silk-spun retreat during the day – this is built in a nook or cranny. They venture out at night to hunt on the ground. Females spin a cocoon to protect their eggs in the summer, and stay close by until they have hatched.

Woodlouse spider *Dysdera crocata*

Arachnida
Length: 10 to 12 mm
Distribution: much of Europe, including Britain.
Observation: all year.

Several very similar species can be found in Europe, but this one is the most widespread. It frequents open, sunny places which generate a warm microclimate. The woodlouse spider has eight oval-shaped eyes on its forehead and powerful hooks which open horizontally, making it perfectly adapted for night-time hunting. It forages over the ground and litter in search of its woodlouse prey and, once captured, it punctures their protective carapace. During the day, it takes refuge in a silk retreat spun in a sheltered spot.

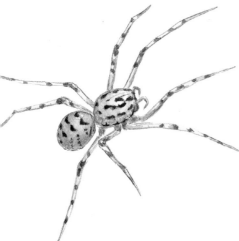

Spitting Spider *Scytodes thoracica*

Arachnida
Length: 4 to 6 mm
Distribution: throughout almost all of Europe.
Observation: June to September.

The Spitting Spider has a distinctive cephalothorax which is almost as big as its abdomen. Both segments are yellowish with parallel black markings. In southern parts of its range it frequents rocks and cracks, while in more northerly regions it is found mainly in houses.

Hunting at night with slow movements, this spider captures prey such as flies by projecting a sticky jet which is produced by the venom glands. The prey is immobilised using a series of fine threads and the spider delivers a final fatal bite.

Garden Spider *Araneus diadematus*

Arachnida
Length: 8 to 12 mm
Distribution: all of Britain and Europe.
Observation: August to November.

The most common species of orb-web spider, the Garden Spider can be found in all habitats with tall-growing vegetation, including gardens, woodlands, hedgerows and heaths. Its web, which can attain a diameter of 40 cm, fastens onto vegetation at a height of 1.5 to 2.5 m from the ground. Webs can be easily spotted during the day but it is interesting to visit them at dusk, which is the best time to watch the spider spinning or repairing its web. Webs are very fragile and are frequently rebuilt at the end of the afternoon or the beginning of the evening.

Wasp Spider *Argiope bruennichi*

Arachnida
Length: 4 to 15 mm
Distribution: mainly central and southern Europe,
occurs locally in south-east England.
Observation: August to October.

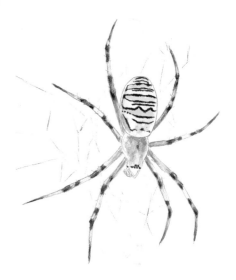

With its black and yellow striped abdomen, the
Wasp Spider is easy to identify. A sun worshipper,
it is found mainly in hot, open places, especially
in the northern part of its range. Its web, which is
crossed by a silken zigzag, is built in tall grass close
to the ground and is an ideal trap for grasshoppers
and crickets.

In autumn the Wasp Spider builds a complex
cocoon in low-growing vegetation which is basically
an upturned balloon that protects its eggs. It
generally spins the cocoon only at night, although
this arduous task sometimes continues into the
early hours of the morning.

CRUSTACEANS

Rough Woodlouse *Porcellio scaber*

Isopoda
Length: up to 17 mm
Distribution: throughout Europe.
Observation: all year.

Woodlice, like all crustaceans, breathe through
gills. Therefore, they inhabit damp areas and are
active predominantly at night to avoid fatal
dehydration. Many very common species, including
this one, which is one of the biggest, live in close
proximity to humans and frequently enter houses.
As soon as the humidity rises at dusk, or even
during the day when the weather is wet, woodlice
leave their shelters beneath stones or in cracks and

crevices to explore areas of moss and other low
vegetation in order to feed on organic debris from
plants and animals.

MYRIAPODS

Stone Centipede *Lithobius forficatus*

Chilopoda
Length: 2 to 3 cm
Distribution: almost all of Europe.
Observation: all year.

With a flat body and legs on its sides, this centipede is adapted to living and hunting in cracks in rocks, under bark or in other confined spaces. It is common in many habitats, particularly gardens and woodland.

The Stone Centipede hunts on the ground or in leaf litter at night, where it searches for insects and their larvae, worms and other invertebrates. It can get into the smallest of cracks and easily extract itself from the narrowest cul-de-sacs thanks to its capacity for moving backwards.

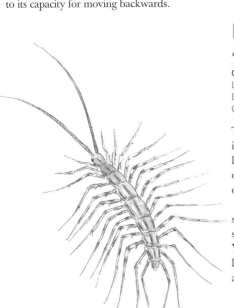

House Centipede
Scutigera coleoptrata

Chilopoda
Length: up to 2.5 cm
Distribution: central and southern Europe.
Observation: all year.

This centipede is found in damp and rocky places in southern Europe. Elsewhere in its range, it only lives in buildings. Unable to cope with the dry air of central or electric heating, it is found mainly in old, damp houses.

The carnivorous House Centipede preys on small insects. It takes refuge in a crack or under a stone during the day, only coming out in the dark. You can sometimes surprise one by turning on a light, while occasionally they become trapped in a bath or sink.

Megarian Banded
Centipede *Scolopendra cingulata*

Chilopoda
Length: up to 10 cm
Distribution: southern Europe.
Observation: throughout the year.

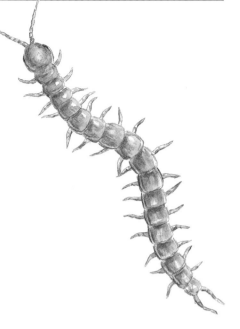

Europe's largest centipede is confined to regions bordering the Mediterranean. It favours open habitat such as scrub where it lives on the ground or under stones, hiding in a burrow during the day. The Megarian Banded Centipede is a formidable predator that only hunts at night. Although not particularly choosy, its preferred prey is nocturnal invertebrates such as spiders, cockroaches and especially snails. Its bite can be very painful to humans.

Millipede *Tachypodoiulus albipes*

Diplopoda
Length: 2 to 3 cm
Distribution: central and northern Europe.
Observation: March to November.

Many species of millipedes are encountered in southern Europe – this is the most northerly. It is only found in mountainous regions in the south of its range and can be found at an altitude of up to 3,000 m. It lives in the litter of dead leaves in forests and hedgerows, preferring cold, damp places with a northerly aspect, although it avoids ploughed fields even if they are covered in plant matter.

Millipedes are essentially active at night, when they roam the ground in search of live or dead vegetation. If disturbed they coil up into a spiral.

Pill Millipede *Glomeris marginata*

Diplopoda
Length: 7 to 20 mm
Distribution: almost all of Europe.
Observation: all year.

This species resembles a woodlouse and also rolls into a ball in the event of danger. The Pill Millipede has much better resistance to dry conditions than woodlice, although it also lives on the ground and feeds on decomposing organic vegetable matter and mushroom mycelium.

This very common millipede frequently ventures out during the day and is commonly observed as soon as dusk falls. The female turns over onto her back to lay eggs, which are covered in excrement to protect them before they are put on the ground.

INSECTS

Dusk Hawker *Boyeria irene*

Odonata
Wingspan: 8 to 9 cm
Distribution: France, Iberia, Italy and locally in Switzerland.
Observation: June to September.

Dragonflies are primarily daytime insects although a few species, particularly hawkers, are active at dusk. The Dusk Hawker is the most nocturnal of these, often flying in total darkness and sometimes even falling prey to bats. It sometimes comes towards artificial lights, but other more diurnal dragonflies are also occasionally attracted in this way.

Although the larva is associated with calm, shaded rivers, the adults use their powerful flight to distance themselves from their place of origin and can be found hunting in a range of habitats, even in towns and gardens.

Green Drake Mayfly *Ephemera danica*

Ephemeroptera
Length: 1.5 to 2.5 cm
Distribution: throughout almost all of Europe
Observation: May and June.

This is one of three very closely related species of mayflies that live around rivers and lakes with sandy or silt bottoms in which their larva can dig tunnels. They have become much less numerous in recent times, perhaps due to their sensitivity to water quality and the loss of pristine habitat.

Mayflies are famous for their very short lives. They generally emerge en masse over a short period of about 10 days at a particular site. The peak period for the adults to fly is in the evening and at the beginning of the night. They devote their few hours of life to mating and egg-laying, and they are attracted to artificial light.

Oak Bush-cricket *Meconema thalassinum*

Orthoptera
Length: 1.2 to 1.5 cm
Distribution: throughout Europe.
Observation: late July until November.

Although common, this species is difficult to observe because it lives exclusively in trees, most notably oak and lime. It hunts at night, preying on invertebrates, especially aphids and small caterpillars.

The male cannot make a typical cricket's churring song on its own, so instead attracts females in the breeding season by drumming out a sound on a leaf or other surface with one of its back legs. This soft humming *tr-tr-tr-trrr-trrr* is barely audible to humans and can only be heard from a distance of 1 m or less.

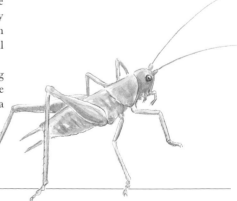

Common Green Bush-cricket *Phaneroptera nana*

Orthoptera
Length: 1.3 to 1.5 cm
Distribution: southern Europe.
Observation: August to October.

This vegetarian cricket feeds on the leaves of a variety of tree species. It frequents woodlands and scrub, usually near a river or wetland, and sings mostly at night, emitting a series of high-frequency rattling sounds with each being distinct from the next. The closely related Sickle-bearing Bush-cricket is slightly bigger at up to 1.8 cm long and is widespread in central Europe. Its modest song is a succession of *tsb* notes that are given at regular intervals.

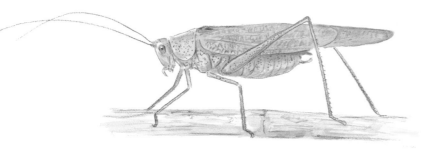

Large Conehead *Ruspolia nitidula*

Orthoptera
Length: 2 to 3 cm
Distribution: central and southern Europe.
Observation: August to October.

This large species is up to 5 cm long if its protruding elytrons are included. It frequents open, grassy habitats such as damp meadows and lawns. It is apparently an omnivore, with its diet including both small insects and grass seeds. This cricket has exclusively nocturnal habits. The males give a long, loud song which runs over into ultrasounds. Shorter, more piercing sounds can be distinguished at regular intervals from amid a very loud and high-pitched continuous vibrating call.

Great Green Bush-cricket *Tettigonia viridissima*

Orthoptera
Length: 3 to 4 cm
Distribution: throughout Europe.
Observation: July to October.

Up to 5 cm in length including its elytrons (wing cases), this large bush-cricket hunts all sorts of insects in both tall grass and tree foliage. It is active during the day as well as at night, and its song can be heard from the afternoon. Singing is more frequent after dusk and can continue all night if the temperature remains above 12°C. Males situated in trees, where the atmosphere cools down more slowly, sing for longer periods of time than those in tall grass. The song, which is audible at a distance of up to 50 m, is loud and quite high-pitched but fitful, and is composed of three to four vibrations per second.

Saddle-backed Bush-cricket
Ephipigger ephipigger

Orthoptera
Length: 2 to 3 cm
Distribution: central and southern Europe.
Observation: July to October.

This species occurs as far north as Belgium and is found in glades, woodland borders and especially vineyards in the northern parts of its range. Unfortunately, its population and range have decreased due to the substantial use of pesticides in wine-making in recent times. It feeds on vegetation and small insects.

The male sings a characteristic two-note *tsee-chipp* both day and night. The female replies to the call of the male and can also sing if held in the hand.

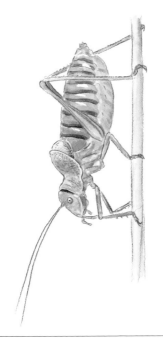

Dark Bush-cricket *Pholidoptera grisoaptera*

Orthoptera
Length: 1.3 to 1.8 cm
Distribution: throughout Europe.
Observation: June to November.

This common cricket prefers forest glades and borders, but can also be found in bushes. It hunts small insects, especially aphids and caterpillars, and complements this diet by devouring the foliage of small plants.

The male sings during the day and at dusk until nightfall. Its regular song is made up of a piercing phrase repeated roughly every four seconds. Normally this is composed of three *tssi tssi tssi* notes, but in very hot weather it is condensed into a short, piercing *tssritt*.

Mole Cricket *Gryllotalpa gryllotalpa*

Orthoptera
Length: 4 to 5 cm
Distribution: throughout Europe.
Observation: almost all year.

Mole Crickets live in the ground, digging tunnels to hunt for the small invertebrates which make up a significant proportion of their diet. These spectacular insects prefer light but damp soil and consequently are most frequently found in valleys.

Considered a pest in some areas because they can cause crop damage, they have seriously declined in many areas due to persecution, although they are protected in Britain.

Males sing a long, piercing, buzzing song from the entrance to their tunnel, particularly at dusk in May and June. Mole Crickets are nocturnal and they are proficient flyers, occasionally coming to artificial light.

Wingless House-cricket *Gryllomorpha dalmatina*

Orthoptera
Length: 1.5 to 2 cm
Distribution: southern Europe.
Observation: August to December.

Remarkable for its extremely long antenna which project the length of its body, this large cricket originally lived in caves and rock cavities, which provide the cool, damp conditions it requires. However, it has adapted to man-made constructions and is now frequently found in cellars and old buildings.

This nocturnal cricket generally stirs when night falls, although occasionally it can be seen in the afternoon when the weather is damp and overcast with little light. As it does not possess wings it does not sing, so the male attracts the female by striking its abdomen against the ground.

European Field-cricket *Gryllus campestris*

Orthoptera
Length: 2 to 2.5 cm
Distribution: throughout Europe.
Observation: May to July.

Although still very common in meadows, glades and crops in the south, the vegetarian European Field-cricket has declined considerably in the northern parts of its range. In the autumn, mature larva select a sunny spot to dig a network of tunnels in which to spend the winter. Males sing in front of the tunnel entrance during the day and at the beginning of the night. Their song is a loud, monotonous *cree cree cree* made up of a single note which is endlessly repeated at a rate of about three times per second. As night falls the temperature drops, causing the tempo of the song to slow down gradually until the cricket becomes silent.

Wood Cricket *Nemobius sylvestris*

Orthoptera
Length: 1 cm
Distribution: all of Europe.
Observation: June to November.

This cricket inhabits leaf litter in woodlands and their borders, scrub and hedgerows. The male emits a rather loud, regular and harmonious song both day and night. The sound is amplified by the presence of several singing males, making it difficult to locate individuals. It is quite a deep, gentle *ruurrr* lasting between a quarter of a second and two seconds depending on the temperature. Calls are interrupted by short silences and repeated indefinitely at a rhythm of about 15 times per minute. This is the species of cricket that sings latest into the autumn.

Italian Cricket *Oecanthus pellucens*

Orthoptera
Length: 1 to 1.5 cm
Distribution: central and southern Europe.
Observation: July to October.

This nocturnal cricket inhabits bushes and long grass, where its light brown colouration provides good camouflage during the daytime. Originating from the Mediterranean, it has spread as far north as Belgium, but only frequents open habitats such as urban gardens in this newly colonised territory. The male sings a smooth but powerful song at dusk and night, which is audible from up to 50 m away although the insect is often difficult to locate. The song is an endlessly repeated *tsrruuu*, its rhythm varying according to the temperature. It is only sung in the daytime during overcast weather.

Grasshoppers are essentially diurnal insects. The males sing from late spring but the songs often seem to merge together when several individuals are competing against each other. Although in general grasshoppers become quiet at twilight to be replaced by the genuinely nocturnal crickets and bush-crickets, a few species continue singing into dusk.

Common Field Grasshopper *Chorthippus brunneus*

Orthoptera
Length: 1.5 to 2.5 cm
Distribution: all of Europe.
Observation: June to October.

This grasshopper is very common in open habitat such as meadows, lawns and woodland borders. The male emits a very brief, weak, shrill *bzzt* song, repeated every two to five seconds. Males which are close together compete with one another.

Meadow Grasshopper *Chorthippus parallelus*

Orthoptera
Length: 1.3 to 2.3 cm
Distribution: throughout Europe.
Observation: June to November.

Another common insect that frequents similar but slightly damper habitat to the last species. Its grating song is a *ssree-ssree-ssree-ssre-ssree* lasting one second that is followed by a three-second pause. This tempo diminishes as the temperature drops. The Large Gold Grasshopper (*Chrysochraon dispar*) is less common and prefers even damper conditions. It is the slowest singer of the group, with its one-second *zeezeezee* repeated every five to 10 seconds.

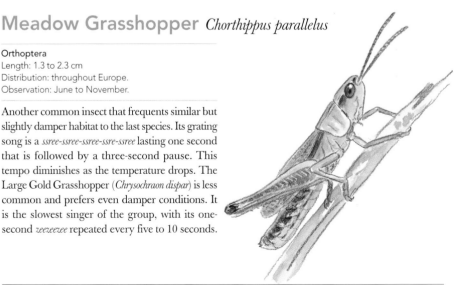

Praying Mantis *Empusa pennata*

Mantoptera
Length: 4.5 to 7 cm
Distribution: southern Europe.
Observation: throughout the year.

Adults can only be found from May until September, although the larvae are active throughout the winter. The adults can be distinguished by the cone-shaped growth above their heads, while the larvae always keep their abdomen raised.

This nocturnal mantis frequents open habitat dotted with bushes, which it uses to hunt its small insect prey. Originally found only around the Mediterranean, the Praying Mantis has spread north, for example along the Atlantic coast of France, although it is not found north of the Loire. It flies frequently and males are often attracted to artificial light.

Dusky Cockroach *Ectobius lapponicus*

Dictyoptera
Length: 1 cm
Distribution: throughout Europe.
Observation: March to November.

Common in woodland litter, this cockroach feeds on dead vegetation. The winged adults can be seen from May to August, while the larvae, which take a year to develop, can be encountered in all seasons except winter.

About 15 other species of cockroach are native to western Europe, with several being confined to the Mediterranean region. All live in similar natural habitats. House cockroaches, which are the cause of serious health problems, are tropical species that are incapable of surviving outside in our climate. All cockroaches are active at night and hide during the day.

Stick Insect *Clonopsis gallica*

Phasmoptera
Length: up to 7 cm
Distribution: southern and central Europe.
Observation : May to October

Half a dozen species of phasmids live in southern Europe and all share a similar twig-like appearance. This stick insect is the most northerly species as its range reaches as far as France's Loire Valley.

The insect remains still during the daytime and is hard to detect as it merges so perfectly into the vegetation. It becomes active at night, grazing on bushes, walking slowly or swaying on its gangly legs. The best way to spot them is with the aid of a lamp, searching at eye level in the tops of bushes, especially brambles, or in long grass.

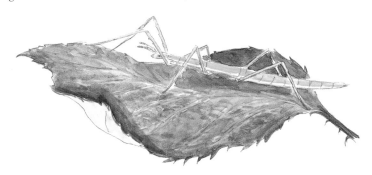

European Earwig *Forficula auricularia*

Dermaptera
Length: 1 to 2.5 cm
Distribution: throughout Europe.
Observation: March to November.

becomes active at dusk, exploring its surroundings in search of suitable plants to eat or small inverebrate prey. It also visits flowers in search of aphids.

Closely associated with humans, this earwig is common in gardens, crops, hedgerows and meadows. It is less so in woodlands, heaths and other more natural environments.

The earwig, with its flat body, easily slips into cracks where it can hide during the day. Being an excellent climber it also colonises tree foliage. It

Although cicadas are diurnal insects, they continue to sing at dusk and sometimes well into the night if the temperature is high. During the day they are easily frightened if you try to approach them. It is easier to find them at night by inspecting vegetation with a torch. There are about 20 species in Europe, with the following two being among the most common.

Red Cicada *Tibicina haematodes*

Homoptera
Length: 3 to 3.5 cm
Distribution: southern and central Europe.
Observation: June to September.

This species is abundant around the Mediterranean, but can also be found further north. Its song is made up of two to five short, successive signals in the space of a few tenths of a second, followed by a loud, continuous chirruping which lasts for several seconds. It is often to be found on a branch or leaf low down in a fruit tree.

Common Cicada *Lyristes plebejus*

Homoptera
Length: 3 to 4 cm
Distribution: southern Europe.
Observation: June to September.

The largest European cicada is confined to the Mediterranean region. Very common, its song is a long, monotonous chirruping. It favours pine trees, especially umbrella pines, from which it extracts the sap, and it can often be found perched on the branches, either close to the ground or several metres up.

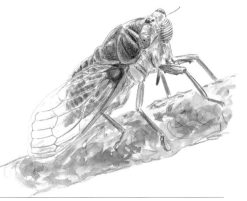

Masked Hunter *Reduvius personnatus*

Heteroptera
Length: 15 to 18 mm
Distribution: almost all of Europe.
Observation: throughout the year.

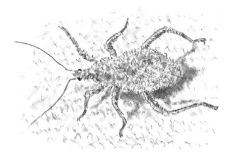

In southern Europe this assassin bug frequents hollow trees where it hunts small prey, dispatching them by sucking out their body fluids. It has a painful sting if you try to pick it up. Further north, it can be found in old buildings, attics and barns. It takes refuge in cracks during the day to avoid the light, coming out at dusk to hunt its prey. The larva coats itself with a sticky liquid that quickly becomes covered in dust that renders it very hard to locate on the ground, apart from when it moves.

Violet Ground-beetle *Carabus violaceus*

Coleoptera
Length: 2 to 4 cm
Distribution: throughout almost all of Europe.
Observation: April to September.

Ground-beetles are large predatory insects that often boast an attractive metallic sheen. There are about 100 species living in different habitats in Europe. This one, which has several subspecies, can be found in gardens and woodlands and on farmland.

Like most of its relatives the Violet Ground-beetle leaves its shelter at dusk to hunt at night. Ground-beetles very rarely fly and usually remaining on the ground or in low vegetation. Certain species which are specialised in snail-hunting can be more easily located after a summer shower.

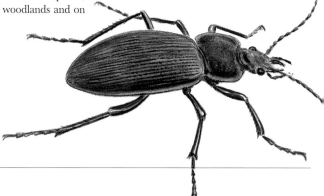

Ground-beetle *Harpalus affinis*

Coleoptera
Length: 9 to 12 mm
Distribution: throughout almost all of Europe.
Observation: all year.

Hundreds of species of ground-beetle can be found in Europe and this species is one of the most common. All live principally on the ground, where different types have evolved to become carnivores, scavengers or vegetarians.

Harpalus affinis can be most easily found at dusk, when it can be seen running over the ground in gardens, crops and meadows, and occasionally in woods. The bulk of its diet is made up of seeds. Other carnivorous ground-beetles tend to be more nocturnal. Adults live for a long time and survive the winter by remaining active.

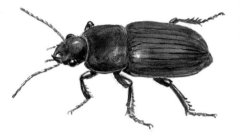

Rove-beetle *Ocypus ater*

Coleoptera
Length: 2 to 3 cm
Distribution: most of Europe, but not Britain.
Observation: March to October.

More than 2,000 species of rove-beetles, or staphylinids, live in Europe. This is one of the largest and most frequently encountered by humans as it is at home in gardens, although it also favours meadows and woods. Staphylinids prefer damp habitats such as leaf litter or other types of decomposing organic matter, where they hunt for small invertebrates.

This rove-beetle is active at night as well as during the day, searching for its prey which it captures using its powerful mandibles. If disturbed, it can raise its abdomen and emit a foul-smelling substance. The very similar Devil's Coach-horse (*Ocypus olens*) is found in Britain.

Stag Beetle *Lucanus cervus*

Coleoptera
Length: 3 to 8 cm
Distribution: most of Europe. In Britain most often found in the south.
Observation: May to September.

The Stag Beetle can be found in woodlands, hedgerows, copses and parks which have mature deciduous trees. The larva lives in the rotting wood of a dead trunk or stump – most frequently oak but sometimes other species.

The male flies at dusk with a loud buzzing which is audible from quite a distance. It appears to be attracted by the powerful scent of flowering lime or chestnut trees, although it is incapable of exploiting these as a source food due to the modification of its mouthpiece which is designed specifically to lick the sap that seeps out of trees. The female is much more discreet and very difficult to see.

Rhinoceros Beetle *Oryctes nasicornis*

Coleoptera
Length: 2 to 4 cm
Distribution: most of Europe, but not Britain.
Observation: June to August.

The larva lives in rotting vegetation such as humus in the topsoil or an old rotting stump. Where the Rhinoceros Beetle occurs close to humans it can colonise piles of wood, compost or rotting manure. The horn on the head of the adult male is the reason for the English name of this species, which is becoming increasingly rare.

Adults, which are generally smaller than the Stag Beetle, are active at dusk, fly strongly and are sometimes attracted to artificial light. A great deal of perseverance or luck is required in order to witness the spectacle of two males confronting each other, horn versus horn, in a battle for a female.

Common Cockchafer *Melolontha melolontha*

Coleoptera
Length: 2 to 3 cm
Distribution: throughout almost all of Europe,
including Britain.
Observation: April to June.

Very abundant in the past, the Common Cockchafer has become less common today and is even rare in some places. Its larvae used to ravage crops but it is now a victim of insecticides and modern farming methods. The species feeds on leaves and can be encountered wherever there are deciduous trees. The adult is crepuscular and flies around the trees at dusk, sometimes in large numbers. It is attracted to light and often approaches moth traps. As the night becomes colder, activity levels decrease until eventually the adults remain immobile in the vegetation.

Pine Chafer *Polyphylla fullo*

Coleoptera
Length: 3 to 4 cm
Distribution: central and southern Europe.
Observation: from April to August.

This species is found mainly in coastal areas and valleys, where the larvae feed upon the roots of grasses and other plants that live in sandy soil. It is the largest European chafer and its colouration is fairly typical for members of this family.

Adults shelter in bushes and trees during the day and only become active at dusk, when they eat pine needles, tamarisk leaves and occasionally oak leaves. A strong flyer, it is attracted to artificial light and can sometimes be found around streetlamps.

Common Burying-beetle *Nicrophorus vespillo*

Coleoptera
Length: 1.2 to 2.2 cm
Distribution: throughout Europe.
Observation: May to September.

Several species of burying-beetles can be found in Europe and this is one of the most common. These insects subsist on the corpses of small animals and can be found wherever this resource is plentiful. Occasionally they use fungi or dung as an alternative.

Burying-beetles locate a corpse by smell, thanks to their highly sensitive antenna. They feed on them and use them to raise their young. Sometimes they have to travel long distances to reach a new corpse, so they fly at dusk or just after dark to reduce risk of being caught by a predator.

Dor Beetle *Geotrupes stercorarius*

Coleoptera
Length: 1.5 to 2.5 cm
Distribution: throughout almost all of Europe.
Observation: from April to September

beetles, it reduces the risk of being preyed upon by insectivorous birds by travelling at dusk or during the night.

This is a dung-beetle and, as the name indicates, it recycles the dung of large mammals. This and similar species are abundant within their range so long as veterinary drugs, such as those for worms, do not eliminate them. This species is found throughout the continent (although only in mountains in the south-west) and is one of many in Europe.

It locates fresh dung by smell, is a strong flier and is often attracted to artificial light. Like burying-

Glow-worm *Lampyris noctiluca*

Coleoptera
Length: 1 to 1.8 cm
Distribution: southern and central Europe, locally in Britain.
Observation: May to July.

Feeding on snails, the glow-worm frequents cool, damp places with tall grass such as meadows and the borders of hedgerows and paths. Habitat loss and the use of insecticides have considerably diminished its numbers, particularly in the north of its range.

In calm, mild weather they can be easily seen at nightfall thanks to the spot of greenish light emitted by the tail-end of the abdomen of the wingless female. The females climb up tall grass stems in order to signal their presence to the winged males.

Firefly *Luciola lusitanica*

Coleoptera
Length: 5 to 8 mm
Distribution: southern Europe.
Observation: May to July.

This small beetle frequents scrub, olive groves and open woodlands. It is associated with places that have an abundance of snails, which provide food for both the adults and larvae.

Both sexes have wings but only the male, which is recognisable by its larger head, flies regularly. During these dusk flights it emits short, bright flashes of light to signal its presence to the females, which sit in low vegetation and reply by flashing back. When a male and female are close there is a real lightshow between the two sexes.

Although some species are equipped with a deterrent, most caterpillars are easy prey for birds. Many have reduced this risk by adopting nocturnal habits and hiding away during the day. If you find leaves that show signs of being eaten and cannot locate the culprit, come back again at night and you are much more likely to find the caterpillars. Thousands of species live in Europe and these two illustrate the diversity of survival strategies that can be adopted.

Spurge Hawk-moth caterpillar *Hyles euphorbiae*

Length: up to 8 cm
Distribution: central and southern Europe. A rare visitor to Britain.
Observation: June to October.

This species can be found particularly on fallow land and beside paths and ditches. Eggs are laid on the leaves of euphorbia plants and, once hatched, the caterpillars gorge themselves on the toxic leaves of their host. The caterpillar indicates it unpalatability to birds with its bright colours. Young caterpillars can be seen during the day, whereas older ones only come out at night.

Queen of Spain Fritillary caterpillar *Issoria lathonia*

Length: up to 3.5 cm
Distribution: throughout almost all of Europe. A rare visitor to Britain.
Observation: March to October.

The butterfly frequents grasslands, heaths and gardens where the caterpillar's foodplants grow – these include violet and borage. There are two generations of butterflies each year, so although the flight period is long the insects themselves are not always in evidence. The caterpillar leaves its foodplant during the day and only returns at dusk.

Lime Hawk-moth *Mimas tilae*

Lepidoptera
Wingspan: 6 to 8 cm
Distribution: almost all of Europe, including England
and Wales.
Observation: May to August.

As the caterpillar feeds on the leaves of lime, and occasionally elm or other deciduous trees, this hawk-moth is associated with wooded areas where the caterpillar's favourite foodplants grow. It can also be found in towns and suburbs where lime trees are planted along pathways. Its colouration is very variable, ranging from green to red and passing through yellow and pink.

The adult does not feed and therefore has no need to visit flowers. Active at night, it only takes to the wing to find a mate and its flight is one of the slowest of all hawk-moths. It often comes to artificial light.

Convolvulus Hawk-moth *Agrius convolvuli*

Lepidoptera
Wingspan: 8.5 to 13 cm
Distribution: migrant to Europe, including Britain
where it is more regular further south.
Observation: June to November.

This remarkable African migrant returns to Europe every summer and even reaches Scandinavia in good years. The caterpillar lives on bindweed. Adults roost with their wings closed on a tree trunk or post during the day, their drab colours and markings making them difficult to spot. They become active at dusk, revealing a pink abdomen while vibrating their wings for several minutes to warm up the flight muscles before take-off. They hover like a hummingbird while gathering nectar from flowers with their very long proboscis.

Death's Head Hawk-moth *Acherontia atropos*

Lepidoptera
Wingspan: 8 to 12 cm
Distribution: most of Europe. Scarce migrant to Britain,
mostly in the south and east.
Observation: May to November.

This migrant hawk-moth has a powerful flight and
returns to Europe each year from tropical Africa.
Its appearances are sporadic in the north but it is
more frequent in the south, although less so than
in the past. Its spectacular caterpillar lives on
potatoes and related plants.

Like most hawk-moths it is mainly nocturnal
and feeds on the nectar of flowers, while it is fond
of sweet liquids and sometimes enters beehives. It
often comes to artificial light. It is extremely well
adapted to nocturnal life and even the chrysalis
hatches at night.

Red Underwing *Catocala nupta*

Lepidoptera
Wingspan: 6 to 7 cm
Distribution: most of Europe, including England and Wales.
Observation: July to September.

The caterpillar feeds on willow and poplar and
the moth can be encountered in woodland and
hedgerows, particularly close to rivers and in humid
places.

This moth starts flying at dusk and is often
attracted to the sap seeping out of lesions in trees
and to rotting fruit. It can easily be attracted by
artificial nectar, as can similar related species. During
the day it roosts on a tree trunk and is rarely
noticeable unless it slightly opens its upperwings
to reveal the black and bright red underwings.

Clifden Nonpareil *Catocala fraxini*

Lepidoptera
Wingspan: 6 to 7 cm
Distribution: south and central Europe. Vagrant to Britain.
Observation: August to October.

This species, also known as the Blue Underwing, is more sensitive to the cold than the Red Underwing and can only be found in warmer areas – it favours warmer microclimates in the north of its range. The caterpillar feeds on ash, willow and poplar and the species is present in damp woods and hedges, particularly near wetlands.

The adult remains stationary on tree trunks during the hours of daylight and only flies at night. It is attracted to light and often enters houses. Within its range, if you leave a light on in a room with the window open at night there is a good chance that you will find one on the wall the following morning.

Gypsy Moth *Porthetria dispar*

Lepidoptera
Wingspan: 3.5 to 6 cm
Distribution: throughout most of Britain and Europe.
Observation: July and August.

The caterpillars have catholic tastes and have been recorded feeding on the leaves of more than 400 plant species. When small they can be carried away on the wind thanks to their long hairs. These factors have led to the wide distribution of this species, which broadly corresponds to that of broadleaved forests in Europe. The moth frequents borders, glades and hedgerows, and to a lesser extent crops, parks and gardens.

The grey-brown male is partially diurnal, whereas the whitish female remains still and hidden close to her cocoon during the day, hardly flying and, if so, only at night.

Leopard Moth *Zeuzera pyrina*

Lepidoptera
Wingspan: 4.5 to 5.5 cm
Distribution: throughout Europe, including much of England and Wales.
Observation: July and August

The caterpillar lives in the woody stems of numerous species of trees, including fruit trees and ornamental species. The moth is frequent in woodlands and hedgerows, as well as parks, orchards, gardens and even tree-lined avenues in towns.

During the day the moth remains stationary on a tree trunk, especially favouring chestnut and ash. It flies at dusk and is often attracted to light. Due to the fact that caterpillars live for several years, the populations of adults can fluctuate a great deal from one year to the next.

Buff-tip *Phalera bucephala*

Lepidoptera
Wingspan: 4 to 5.5 cm
Distribution: throughout Britain and Europe.
Observation: May to August

Caterpillars live in colonies on trees such as oak, poplar, willow, birch and beech. The moth is associated with deciduous woods, especially those on river banks, but is also at home in hedgerows, wooded borders, parks and gardens.

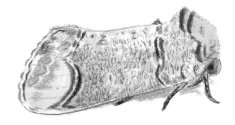

The Buff-tip has a strong and often fatal attraction to lamps which emit a high level of ultraviolet light, and has suffered a high death toll due to street lighting which uses mercury vapour lamps. During the day, the caterpillar is almost invisible, hiding in the vegetation and resembling a fragment of dead wood in order to conceal itself from predators.

Oak Eggar *Lasiocampa quercus*

Lepidoptera
Wingspan: 5 to 7 cm
Distribution: all of Britain and Europe.
Observation: May to July.

As the caterpillar develops on a range of deciduous trees, and even ivy, this moth is associated with broadleaved woods and heaths, and not only to areas with oak trees. It also frequents parks and gardens where it is particularly fond of fruit trees. The dark-brown male can be seen on the wing during the day, often in search of females with its fast zigzag flight. The light brown female is much more discreet and less active, often choosing to wait for the arrival of a male and flying most frequently at dusk. Both sexes come to artificial light.

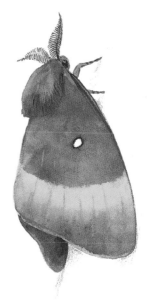

Tau Emperor Moth
Aglia tau

Lepidoptera
Wingspan: 7 to 8 cm
Distribution: central and southern Europe.
Observation: April to July.

The Tau Emperor Moth is closely associated with beech trees and is found predominantly in beech woods. Less common food plants for the caterpillar include oak, birch and other deciduous trees.

The males are mainly diurnal and are easy to observe in flight. They often take to the wing in search of females, particularly early in the morning on warm and sunny days. The females keep a lower profile and tend to hide in vegetation during the day, where they merge in with the dead leaves at the foot of birch trees. They become active at dusk and are attracted to light.

Emperor Moth *Eudia pavonia*

Lepidoptera
Wingspan: 5 to 7 cm
Distribution: throughout almost all of Britain and Europe.
Observation: April and May.

This moth inhabits a broad range of habitats including heathland, scrub, marshes and hedgerows, with its caterpillar developing on herbaceous plants such as heather, meadowsweet or bramble.

The two sexes have very different habits. The male is especially active during the day, searching for females who release a scent to attract them. In contrast the females stay hidden in the undergrowth during daylight to avoid predators; they are difficult to locate and only become active at night.

Giant Peacock Moth *Saturnia pyri*

Lepidoptera
Wingspan: 12.5 to 17 cm
Distribution: central and southern Europe. Occasionally recorded in Britain, perhaps as an introduction.
Observation period: May and June.

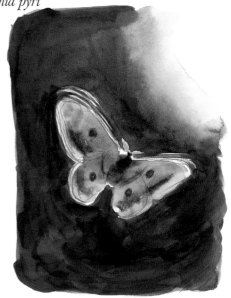

One of the largest European moths. It frequents woods, fallow land, orchards and parks. Its caterpillar lives on a variety of broadleaved trees, especially those of the rosaceae family, which encompasses several fruit trees. It is sensitive to insecticides, which have caused its range to contract greatly.

Both sexes of the Giant Peacock Moth shelter during the day and only emerge at dusk. The male is easier to spot in flight, as it sometimes whirls around streetlamps. The female attracts the male by emitting a scent. Both sexes have a very short lifespan as they do not feed.

Yellow-tail *Euproctis similis*

Lepidoptera
Wingspan: 3 to 4 cm
Distribution: throughout almost all of Britain and Europe
Observation: June to August.

With the caterpillar feeding on a variety of trees and bushes, including hawthorn, oak and willow, this species can be encountered in woodlands, hedgerows and parks. A closely related species, the Brown-tail (*Euproctis chrysorrhoea*), lives in similar habitat and is more common in Europe. It can be distinguished by its brown abdomen end, that of the Yellow-tail being yellow to orange. The Brown-tail is extending its range from south-east England and is considered a pest because the wind-blown caterpillar's hairs can cause a skin rash in humans. Adult Yellow-tails can sometimes be seen during the day, especially egg-laying females. However, these short-lived non-feeding moths fly most frequently at night.

Silver-Y *Autographa gamma*

Lepidoptera
Wingspan: 3.5 to 4 cm
Distribution: throughout Britain and Europe.
Observation: April to November.

One of the most common nocturnal moths in Europe, the Silver-Y is a migrant which can be found all over the continent, although it only maintains its numbers in the north through regular influxes from populations further south.

The Silver-Y can be seen flying or feeding during the day, and is particularly fond of buddleia. However, it is most active at dusk and the beginning of the night, when many individuals can come to gather nectar from flowers, including garden favourites such as phlox and petunia. Although they often alight on the flower, they are capable of gathering nectar in flight, in a similar way to hawk-moths.

Angle Shades
Phlogophora meticulosa

Lepidoptera
Wingspan: 4.5 to 5 cm
Distribution: almost all of Britain and Europe.
Observation: April to October.

The caterpillar feeds on plants as varied as bracken, dandelion, birch and willow, and can be found everywhere from woods and fallow land to crops and gardens, including those in towns and suburbs. When it alights on the ground or on vegetation during the day, its colours and the shape of its wings merge with those of dead leaves. The Angle Shades flies at night when it often approaches artificial light. It is also attracted to sweet liquids and is regularly found at nectar traps. It is particularly common in September, when it frequently gathers nectar from Ivy flowers.

Bright-line Brown-eye *Lacanobia oleracea*

Lepidoptera
Wingspan: 3.5 to 4 cm
Distribution: throughout Britain and Europe.
Observation: May to September.

Nettle, dock, cabbage, asparagus and many other low-growing wild or cultivated plants provide food for the caterpillar. This moth is just as frequently found in saltmarshes and fallow land as it is in crops and gardens, although the damage caused to vegetables is generally minimal.

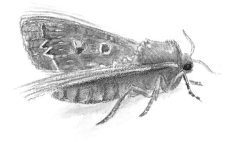

This noctoid moth is completely nocturnal, remaining stationary on a wall or hidden in low-growing vegetation during the day. At nightfall, it comes out to gather nectar from scented flowers, particularly lilac and honeysuckle.

Old Lady *Mormo maura*

Lepidoptera
Wingspan: 5.5 to 6.5 cm
Distribution: throughout almost all of Britain and Europe.
Observation: July to September.

The caterpillar feeds on a large variety of plants, including dock, hawthorn, blackthorn, willow and birch. The moth flies mainly in damp habitats where these host plants grow.

The Old Lady remains motionless in a bush or tree during the day. It occasionally takes refuge in cellars and often shelters in Ivy, which is another favourite food plant for the caterpillar. Rather retiring in its habits, it is only active at night, but can be easily attracted by sweet liquids and nectar traps.

Garden Tiger *Arctia caja*

Lepidoptera
Wingspan: 5 to 7 cm
Distribution: throughout Britain and Europe.
Observation: June to August.

The caterpillar lives on nettles, so this moth frequents a variety of habitats where this host plant grows, from forests to fallow land and from hedgerows to parks and gardens.

Because its bright colours and patterns provide little camouflage, the moth can be found with relative ease during the day as it roosts on a tree trunk or in low-growing vegetation. When disturbed, it defends itself from predators by displaying its bright orange and black underwings, a colour combination that indicates danger. Although it sometimes flies in late afternoon, it is most frequently active at nightfall, when it can be attracted towards light.

Brimstone Moth *Opistograptis luteolata*

Lepidoptera
Wingspan: 3 to 4 cm
Distribution: throughout Britain and Europe.
Observation: May to October.

Woodland, hedgerows, scrub, parks and gardens are the domain of this widespread moth, whose caterpillar feeds on honeysuckle, willow, mountain ash and hawthorn.

The moth can be mistaken for a dead leaf when it is stationary during the day. It becomes active at night and, like all geometrid moths, has a limp flight and rarely covers great distances. It is often attracted to light, fluttering around in the manner characteristic of geometrids.

Peppered Moth *Biston betularia*

Lepidoptera
Wingspan: 4.5 to 5.5 cm
Distribution: most of Britain and Europe.
Observation: May to August.

This geometrid moth frequents deciduous forests, especially those with damp or marshy ground, as well as parks and gardens. The greyish coloured moth remains motionless on bark during the day, where it can easily go unnoticed. It is famous for having been used to help prove Darwin's theories in the 19th century, when natural selection caused many individuals in the industrial towns of England to become almost black in order to remain camouflaged against the soot-blackened tree trunks and avoid predation from insectivorous birds. The moth is active from dusk and during the night, with both sexes coming to light.

Caddisfly *Phryganea grandis*

Trichoptera
Length: 1.5 to 2 cm
Distribution: central and southern Europe.
Observation: May to July.

The larvae live at the bottom of stagnant or slow-moving waters covered in aquatic vegetation, and are renowned for their protective cases. The adults rarely travel far from this wetland habitat and, as the mouthpiece is poorly developed, they feed only on a small amount of liquid during their very short lives.

During the day caddisflies remain stationary on a leaf, stem or stone, or in a crack, with head down and antenna pointing forward. They become active at dusk, sometimes flying in groups over the water and vegetation, and they are attracted to light.

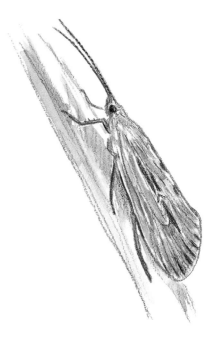

Mosquito *Culex pipiens*

Diptera
Length: 4 to 5 mm
Distribution: throughout Europe.
Observation: all year.

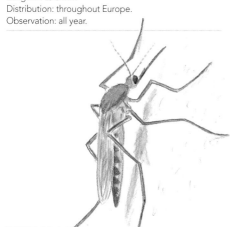

As the aquatic larvae are capable of developing in even an old tin can filled with rainwater, these insects can be found everywhere. Several species which bite humans occur in Europe, this one being the most common. They are particularly abundant in damp and humid environments and can be a nuisance on a night-time outing in such places.

Only the female bites, in order to obtain blood to mature her eggs. They locate their warm-blooded victims by detecting the carbon dioxide released during breathing. Essentially nocturnal, they are attracted to light. Adults can be encountered year round, but are less active in winter.

Mantis-fly *Mantispa styriaca*

Nevroptera
Size: 1 to 2 cm
Distribution: central and southern Europe.
Observation: May to September.

Looking like a miniature praying mantis, mantis-flies are more retiring and less well known than their namesake. The two insects are unrelated but share similar habitat, favouring warm and sheltered bushy areas with plenty of sunlight.

The mantis-fly is an efficient predator, hunting small insects up to its own size, especially flies. It hunts in trees and bushes during the day and at night. It keeps a low profile and it is difficult to locate. Several species of mantis-fly can be found in regions bordering the Mediterranean, with cork-oak being a preferred habitat.

Green Lacewing *Chrysoperla carnea*

Nevroptera
Size: 7 to 12 mm
Distribution: throughout almost all of Europe
Observation: April to October.

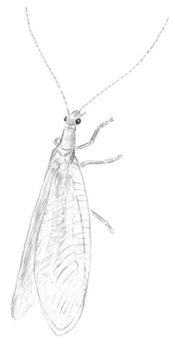

This insect occupies all habitats with abundant vegetation, from low-growing plants to tall trees, as long as they harbour good numbers of its favourite prey, aphids. Green Lacewing is one of about 30 closely related species that exist in Europe. Like most nevroptera it has a lethargic, undulating flight but can cover quite long distances. It is often attracted to light, especially at dusk and the beginning of the night. Lacewings are able to detect ultrasound, enabling them to escape more easily from bats.

Ribbon-tailed Lacewing *Nemoptera bipennis*

Nevroptera
Wingspan: 5 to 7 cm
Distribution: Spain and Portugal. Similar species elsewhere in southern Europe.
Observation: April to August.

This species occurs in Iberia, where it frequents open, rocky scrubland as well as forests, especially pinewoods. The adult can be found in long grass and bushes. The larva, with its exceptionally long neck, hunts on the ground. Other similar species in the southern and eastern Mediterranean boast similar remarkably long hind-wings which trail behind like a tail.

These insects can be recognised by their lethargic, undulating flight at dusk. They sometimes occur in groups and are often attracted to artificial light.

Ant-lion *Palpares libelluloides*

Nevroptera
Wingspan: 10 to 12 cm
Distribution: southern Europe.
Observation: May to September.

More than 20 types of ant-lion occur in Europe. The larvae of certain species excavate a distinctive funnel-shaped trap in sandy soil to catch insect prey. All frequent open, warm, dry environments. This species is particularly common in Mediterranean scrubland.

Adults are most frequently found roosting in tall grass. Their flight is slow and lethargic, and if disturbed an individual will usually alight just a few metres away. Most ant-lions are crepuscular or nocturnal. This species is largely diurnal, but all species come to light.

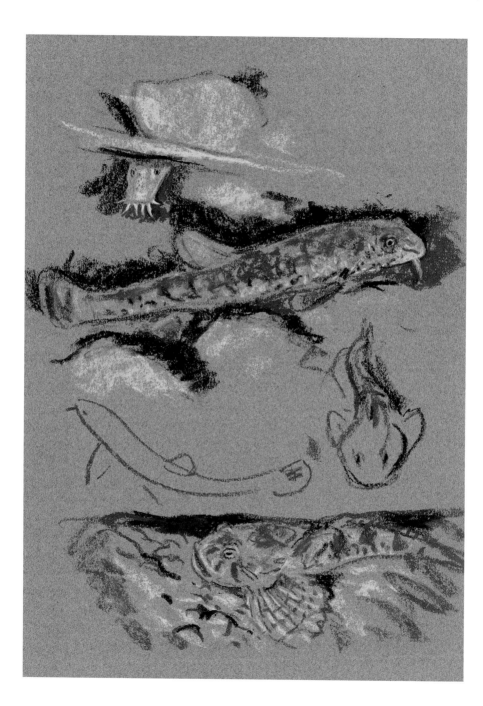

Aquatic invertebrates and fish

The aquatic world is so distinct that it warrants a separate section in this book. Water temperature is a good deal less sensitive than air temperature to the nocturnal cooling caused by the temporary disappearance of the sun's energy. Aquatic invertebrates are therefore less prone to nocturnal cooling and can remain active for a longer time.

Furthermore, most aquatic creatures, whether they are diurnal or nocturnal, are attracted to artificial light, at least once they have got over their initial instinct to flee. It is difficult to watch aquatic fauna at night using only natural light, even with bright moonlight, due to reflections and suspended matter. Therefore the use of a torch is essential.

Of course, space does not permit an exhaustive review of all the species that could be encountered. So as with the chapter on invertebrates a small number of species, often the biggest or most spectacular, have been selected as examples of entire families. This selection provides a concise introduction to the discovery of nocturnal life in ponds or rivers, which can be complemented by referring to more extensive works by those wishing to broaden their knowledge and understanding.

Stone Loach (above) and Bullhead

ANNELIDS

Horse Leech *Haemopis sanguisuga*

Acheta
Length: up to 10 cm, sometimes larger
Distribution: throughout almost all of Europe
Observation period: all year

Living in shallow water close to the shores of lakes and ponds, this common leech has a very varied diet. Although it sucks blood from livestock that comes to water to drink – hence its common name – or from humans, it feeds predominantly on molluscs and other small invertebrates, from which it sucks the body fluids. It sometimes buries itself in the ground while searching for earthworms. The Horse Leech, which can grow to a spectacular size, is found most frequently under stones, but can sometimes be seen searching for prey in open water.

MOLLUSCS

Great Pond Snail *Limnaea stagnalis*

Gastropod
Length: up to 6 cm
Distribution: throughout Europe.
Observation: all year.

Great Pond Snails frequent still or slow-moving bodies of water, providing they are rich in underwater vegetation. They graze by scraping the surface of plants with their tongue and if necessary can tolerate slightly salty water.

Moving at a similar speed to their terrestrial cousins, they crawl across the stalks and leaves of the plants upon which they feed. They are easy to observe by day or night year-round and they possess lungs rather than gills.

CRUSTACEANS

White-clawed Crayfish *Austropotamobius pallipes*

Decapoda
Length: up to 13 cm
Distribution: western Europe, including Britain.
Observation: all year.

Once very common, this freshwater crayfish has been a victim of water pollution and a disease imported with similar species introduced from America. Today the indigenous species has become rare and its decline has accelerated because it tends to be replaced by introduced species which have a higher tolerance to pollution and disease.

All crayfish, whether carnivorous, vegetarian or scavengers, hide under stones or in holes. Mainly nocturnal, they hunt for food by walking rather than swimming.

Freshwater shrimp
Gammarus lacustris

Amphipoda
Length: 1 to 2 cm
Distribution: throughout Europe.
Observation: all year.

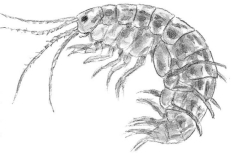

Common in still or flowing water, freshwater shrimps are represented by several closely related species in Europe, this being one of the most common. They live in ponds, ditches, lakes, streams and shallow rivers. These natural refuse collectors consume organic detritus.

Freshwater shrimps have an unusual way of swimming, on their side, contracting and stretching the body. They can sometimes be found in fast-flowing water, where they shelter under stones or in vegetation. They are much more easy to spot in still water.

INSECTS

The odonata are divided into two groups: dragonflies keep their wings at right angles to their bodies at rest while damselflies fold them along their bodies. The aquatic larvae of dragonflies and damselflies are also distinctive. Attributing these nymphs to a species usually requires expert knowledge, but the two groups can be easily distinguished. These larvae are ferocious carnivores and are found in underwater vegetation. They walk on the river or lake bed or swim in open water, and are frequently attracted to light.

Damselfly larva *Zygoptera* sp.

Odonata
Length: up to 3 cm
Distribution: throughout Europe.
Observation: almost all year.

The damselfly larva is thin and spindly with an abdomen finishing in three flat appendages which form a leaf-shape. These are the gills which it uses to breathe and also to move because they act like tail-fins.

Dragonfly larva *Anisoptera* sp.

Odonate
Length: up to 5 cm
Distribution: throughout Europe.
Observation: almost all year.

Wide and squat, the dragonfly larva has an abdomen finishing in four short appendages, which do not serve as gills. It moves by walking, or by expelling air from its stomach which acts as a sort of jet to propel it rapidly forwards

Water-scorpion *Nepa cinerea*

Heteroptera
Length: 1.5 to 2 cm
Distribution: throughout almost all of Europe
Observation: all year.

The water-scorpion can be found in ponds, lakes and slow-flowing rivers, walking over silty beds or clinging to vegetation. Breathing through its long, narrow tail appendage, it stays in shallow water to avoid drowning.

The water-scorpion uses its pincer-like front legs to hunt and sometimes covers itself with silt as camouflage. It is not very mobile and has a blackish colouring that makes it difficult to locate, although it can be attracted to light. Despite having wings, the water-scorpion does not fly, although it does sometimes come out from the water onto the bank.

Water Stick Insect
Ranatra linearis

Heteroptera
Length: 3 to 4 cm
Distribution: throughout almost all of Europe
Observation: all year.

Like the water-scorpion, the Water Stick Insect hunts using its long pincer-like front legs and breathes through a long tube located at the end of its body. It frequents similar habitats, but prefers to remain in vegetation. Its long, spindly body and pale brown colouring help it to merge in with its background and it can be easily mistaken for a twig. The water stick-insect walks slowly and can be attracted to artificial light. Out of water they can fly over short distances, especially during the day, and they occasionally stridulate like crickets.

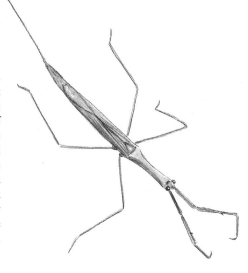

Common Water-boatman *Notonecta glauca*

Heteroptera
Length: 1.5 cm
Distribution: throughout Europe.
Observation: all year.

With its hind-legs transformed into oars, the Common Water-boatman swims on its back in open water. Several closely related species exist in Europe. They stay close to the surface, breathing air accumulated under their wings through the end of their abdomen. They are always ready to pounce on prey that passes close by or gets stuck on the surface.

The Common Water-boatman is easily attracted to light. It flies well, with a dull buzzing sound, moving from one stretch of water to another essentially at night. It is generally one of the first species to colonize a new pond.

Common Pondskater *Gerris lacustris*

Heteroptera
Length: 8 to 15 mm
Distribution: throughout Europe.
Observation: almost all year.

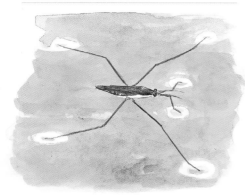

About 10 different species of pondskaters glide across the surface of fresh and briny waters in Europe. This species is the most common and is confined to still or slow-moving waters. Other species skate in jumps and are capable of crossing quite strong currents. Pondskaters hunt small prey which falls onto the surface of the water.

They are easy to observe in a torch beam and often live in small colonies which can be found through carefully inspecting the surface of the water. They fly especially at night in search of new watercourses to colonise, emitting a buzzing sound in the process.

Whirligig beetle
Gyrinus natator

Coleoptera
Length: 5 to 7 mm
Distribution: throughout Europe.
Observation: March to October.

Several closely related species of whirligig beetles can be found in Europe, and all wheel around swimming energetically on the surface of the water. Always found in groups, they hunt small prey which has become stuck on the water's surface and sometimes scavenge on corpses. They can dive to avoid danger.

Whirligig beetles are voracious predators that live in still or slow-flowing waters. They can be easily observed in artificial light. Mainly in autumn they fly at night in search of new places to colonise.

Great Diving-beetle *Dytiscus marginalis*

Coleoptera
Length: 3 to 3.5 cm
Distribution: throughout Europe.
Observation: almost all year.

This very common insect frequents still or slow-moving waters where vegetation is abundant. It is a carnivore that hunts prey such as small tadpoles and young fish while swimming, and it resurfaces from time to time to renew the supply of air accumulated under its elytrons.

The Great Diving-beetle can be easily observed in torchlight. It flies well, most often at night when it colonises new ponds. Occasionally it mistakenly lands on wet tarmac or a dark plastic tarpaulin. In such cases it can be seen walking awkwardly and incapable of returning to the air.

Great Silver Water-beetle *Hydrophilus piceus*

Coleoptera
Length: 4 to 5 cm
Distribution: throughout almost all of Europe.
Observation: March to November.

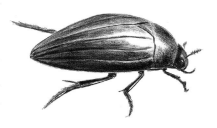

Superficially quite similar to the Great Diving-beetle, the Great Silver Water-beetle is much larger but not such a fearsome predator. It is vegetarian when adult and carnivorous only at the larval stage. It does not swim as strongly as the diving-beetle and renews its air supply using its antenna, coming to the surface head-first to breathe rather than with its head underwater like the diving-beetle. This is an easy way to distinguish the two species.

This insect may require a little patience to observe in the torchlight. It inhabits still or slow-flowing waters with plentiful vegetation. It is a good flyer, moving from one site to another if conditions no longer suit it.

ARACHNIDS

Water Spider *Argyroneta aquatica*

Arachnida
Length: 8 to 15 mm
Distribution: throughout Europe.
Observation: March to November.

Once much more common, the only European aquatic spider has become scarce in recent times. It builds a silk bell in the underwater vegetation in still or slow-flowing waters. Here it creates an refuge by accumulating air bubbles which have been fastened onto its hairs.

The Water Spider shelters in its bell during the day. It leaves at night to hunt small invertebrates, walking on the bed of the stream or lake. It swims on its back, rowing with its legs and appears to shine when moving in the water due to the air for breathing accumulated in its hairs.

FISH

Eel *Anguilla anguilla*

Anguilliforme
Length: Up to 1.5 m
Distribution: western and central Europe.
Observation: throughout the year.

The Eel is found as frequently in estuaries as it is in rivers, lakes and ponds. It leads a sedentary existence until it embarks upon a long migration at the end of its life in order to reproduce at sea. The Eel leaves its daytime shelter at nightfall to hunt insects, crustaceans, fish and amphibians, and to scavenge at corpses. This fish is capable of leaving the water and slithering across land like a snake, especially during damp or rainy nights or after a heavy storm.

Allis Shad *Alosa alosa*

Clupeiformes
Length: 35 to 70 cm
Distribution: western Europe.
Observation: May to July.

The Allis Shad is primarily a coastal fish that swims up rivers to breed in spring. Once common, its populations have decreased drastically due to river development.

At nightfall in June and July, the male and female swim round in circles for a few seconds producing a characteristic bubbling that sounds rather like a dog snorting. Each of these events provokes a ripple on the surface of the water which is visible on a still night. This activity generally reaches a peak during the middle of the night, between two and three o'clock in the morning.

Burbot *Lota lota*

Gadiformes
Length: from 25 cm to 80 m
Distribution: central and northern Europe.
Observation: all year.

This freshwater relative of the cod lives in the cold running waters of rivers, but also in lakes and sometimes brackish lagoons. It is a deep-water species that hunts insects, crustaceans and small fish. During the day the Burbot takes refuge under a large stone, in a hole in a rock or in dense aquatic vegetation. It emerges at dusk and spends much of the night actively searching for food. This species can endure cold water and is active throughout the year, even reproducing in winter.

Bullhead *Cottus gobio*

Scorpaneiformes
Length: 10 to 15 cm
Distribution: throughout almost all of Europe
Observation: all year.

The Bullhead requires well-oxygenated water and is usually found in fast-flowing rivers and streams, but occasionally also in lakes. As it does not possess an air-bladder, it lives between or under stones on the riverbed, merging well with the substrate due to its colour and form.

It is active at night, leaving its refuge to hunt small crustaceans, insect larvae and fish fry, and it can even devour small eels. In the spring, the female deposits her eggs in a nest hollowed out by the male, who watches over them until they hatch one month later.

Stone Loach *Nemacheilus barbatulus*

Cypriniformes
Length: 8 to 15 cm
Distribution: throughout almost all of Europe
Observation: all year.

Like the Bullhead, the Stone Loach requires the cool, well-oxygenated water of fast-running rivers or clear lakes. It lives on sand or gravel beds, eating insects, small crustaceans and molluscs.

Hiding under a stone or in vegetation during the day, this fish becomes active at dusk. Sensitive barbels around its mouth help it to locate its food and also to keep it in contact with the riverbed when it swims. Other fish with barbels, such as the sturgeon or catfish, also have nocturnal habits.

Common Carp *Cyprinus carpio*

Cypriniformes
Length: 40 cm to 1 m
Distribution: throughout almost all of Europe
Observation: March to November.

A fish of still or slow-moving waters, the carp prefers silt riverbeds with dense vegetation. Although it lives predominantly in deep water, it frequently comes close to the shore during active periods.

The carp is often active at dusk, when it digs into the riverbed in search of food. At night, it is attracted to artificial light, as are most other diurnal fish. In summer, when the water is warm it can remain active well into the night. Some diurnal fish, such as Roach, Bream and Perch, also become crepuscular or even nocturnal in these conditions.

Amphibians and reptiles

Amphibians could have been covered in the previous chapter since all are aquatic at some point in their life-cycle. However, as all have a significant terrestrial phase in their lives a separate chapter has been devoted to them. A few species of reptiles have been included, these being exceptions in a group which is largely diurnal.

Compared to invertebrates, fish and birds, there are only a small number of species of amphibians and reptiles in Europe. The selection covered brings together some of the most remarkable and the most frequently encountered species.

Among the amphibians, frogs and toads are also remarkable for their mating calls. Males, and occasionally females, produce a chorus at dusk and into the night from early spring into summer. With a little practice it is quite easy to identify the different species present at a particular site without even seeing them. Moonlight is often sufficient for finding amphibians or reptiles, but a torch is usually essential in order to make an accurate identification.

Common Midwife Toads

AMPHIBIANS

Newts lead an amazing double-life. During the late winter and spring breeding season they become diurnal and aquatic. The males become brightly coloured in order to attract a mate, making them much more easy to identify at this time of year. They remain active at night, frequenting more open areas of water and coming readily to torchlight. The rest of the year they mostly live on land are more or less nocturnal. The males lose their bright breeding colours and the young are particularly tricky to identify, while in general all become much more difficult to detect. Newts are carnivorous and feed on a broad range of of invertebrates.

Smooth Newt *Triturus vulgaris*

Urodela
Length: up to 11 cm
Distribution: throughout most of Britain and Europe
Observation: almost all year.

The Smooth Newt prefers shallow, stagnant water with abundant vegetation for breeding. It inhabits marshes, ponds, pools, ditches and also brackish water on the coast. During its terrestrial phase it wanders further from water than most other newt species and can be found in woods, meadows, fields and gardens. This species is generally the first to colonise ponds in towns and cities.

Great Crested Newt *Triturus cristatus*

Urodela
Length: up to 16 cm
Distribution: central and northern Europe, including Britain.
Observation: almost throughout the year.

This newt is the most nocturnal of the species shown here, although during the breeding season, when it is found in still or slow-moving water, it can often be seen in the daytime as well. At other times of year it can be found at a distance of up to 1 km away from its breeding sites and it is particularly fond of woodland, although some individuals spend all year in the water.

Alpine Newt *Triturus alpestris*

Urodela
Length: up to 12 cm
Distribution: central and southern Europe.
Observation: almost all year.

This highly aquatic newt breeds mostly in ponds and pools, but occasionally in calm rivers where fish are absent or scarce. Outside the breeding season it never ventures further than a few hundred metres from the water and always shelters in cool, damp places.

In the Alps it lives at altitudes of up to 2,000 m, and such populations are principally aquatic all year round.

Palmate Newt *Triturus helveticus*

Urodela
Length: up to 9.5 cm
Distribution: western Europe, including Britain.
Observation: almost all year.

Predominantly associated with ponds, pools and slow-moving rivers, this species can also breed in puddles, troughs, ditches and even brackish coastal waters. It is less tolerant of the presence of fish at its breeding sites than the Alpine or Smooth Newts, and the male can be separated from the latter due to a narrow filament at the tip of its tail. Even when not breeding it rarely strays far from water.

Marbled Newt *Triturus marmoratus*

Urodela
Length: up to 16 cm
Distribution: south-west Europe.
Observation: almost all year.

This species is not particularly sensitive to water quality and breeds in a variety of habitats including village ponds, peat bogs, pools and sometimes even troughs.

It is the most terrestrial of all European newts, only coming to water to breed. It can be found in woodlands, grasslands, hedgerows, heaths and scrub.

It is susceptible to being run over by cars, particularly on rainy nights. In winter it shelters in rodent burrows, under the loose bark of trees or in a hollow trunk or branch, sometimes quite high off the ground. It often finds refuge in close proximity to humans, particularly in old cellars or woodpiles.

Fire Salamander *Salamandra salamandra*

Urodela
Length: up to 20 cm
Distribution: much of continental Europe.
Observation: almost all the year.

Difficult to confuse with any other amphibian the Fire Salamander is essentially a forest-dwelling species, although it can also be found in scrub and farmland with hedgerows. It is terrestrial for most of its life but never strays far from water. Its patterning is variable, with some individuals predominantly yellow with black markings or mostly black spotted with red. The species is nocturnal and especially active on rainy nights or following a wet period. At such times they can be commonly seen at roadsides, where many are run over because they are too slow to escape the approaching cars. In winter it shelters in a hole or hollow tree and can sometimes found in woodpiles.

Common Midwife Toad *Alytes obstetricans*

Anura
Length: 4 to 5 cm
Distribution: western Europe, but in Britain only
introduced to a few sites in England.
Observation: March to November.

Exclusively terrestrial at as an adult, the midwife toad frequents woods, quarries, embankments, verges and gardens close to water and is often found living close to humans. The male carries strings of fertilized eggs wrapped around its hind legs.

Its song, which can be heard mainly from May to July, is a succession of shrill, flute-like *too* notes, given every one to three seconds – it is shriller, shorter and carries further than the song of the Yellow-bellied Toad and can be confused with the song of the Scops Owl. The adult midwife toad is crepuscular and nocturnal, hiding under a stone or in a hole during the day.

Yellow bellied Toad *Bombina variegata*

Anura
Length: 4 to 5 cm
Distribution: central and southern Europe.
Observation: March to November.

This small toad is associated with temporary or permanent wetlands which are often man-made. When hibernating on land it never ventures far from its aquatic breeding grounds.

The best time to listen to the male's mating call is in calm, mild weather in May and June, during the day and at the beginning of the night. This call is relatively quiet *bouh* which is repeated every one or two seconds. Always alert, it dives into the water as soon as danger approaches. If it is threatened on land, it arches up to expose the yellow markings on its belly.

Western Spadefoot *Pelobates cultripes*

Anura
Length: 7 to 10 cm
Distribution: Iberia and southern and western France.
Observation: March to June and late August to November.

Burying itself when not active, this rather rare toad frequents open areas with soft ground and water nearby, from coastal dunes to crops and marshland. Its call can be heard mainly in March and April and is often given underwater – it is a *co-co-co* that sounds like the hushed clucking of a chicken and which is only audible at close range. The nocturnal adults only come out two or three hours after sunset on calm nights when the light of the moon is weak or absent and the temperature is above 8°C. The similar Common and Eastern Spadefoots are found in central and eastern Europe.

Parsley Frog *Pelodytes punctatus*

Anura
Length: 4 to 5 cm
Distribution: western continental Europe.
Observation: February to November.

This species prefers open habitat such as meadows, lawns, scrub, open woods, crops and even gardens. It breeds in temporary or permanent areas of water. The call can be heard principally at night in March and April, when the male emits a loud *coe-ak coe-ak* underwater, and the female replies with a more subdued *coo coo*. On land the male emits a squeaky *creeek*. The Parsley Frog hunts at dusk and during the night, jumping to pursue its prey, and it can also climb smooth surfaces and be found perched high in a bush or on a rock.

Natterjack Toad *Bufo calamita*

Anura
Length: 6 to 9 cm
Distribution: western and central Europe.
Very localized in Britain.
Observation: March to November.

This pioneering species is found particularly in open habitat such as dunes, meadows, heaths and quarries, breeding in still water such as ponds and shallow puddles.

The call of the males is audible from April to August and carries well. Beginning just after dark, it is a repetitive, rolling sound that lasts one or two seconds and is similar to the call of a cricket. The adult is nocturnal and hunts insects, worms, slugs and other invertebrates on warm, calm, damp nights. It moves by running in fits and starts rather like a small rodent, and it can also climb.

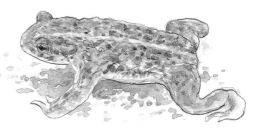

Common Toad *Bufo bufo*

Anura
Length: 7 to 15 cm
Distribution: throughout Britain and Europe.
Observation: February to November.

The Common Toad is terrestrial and occupies a range of habitats including woodland, gardens, meadows, heaths and marshes, as long as there is some permanent water in the vicinity for breeding.

Prior to the breeding season there is a massive migration towards these wetland sites on mild nights. The male gives weak yelping cries, most often in March and April. The Common Toad is essentially a nocturnal hunter, emerging at dusk and roaming the ground in search of its prey.

Green Toad *Bufo viridis*

Anura
Length: 7 to 10 cm
Distribution: central and eastern Europe.
Observation: February to October

Favouring dry, open habitats, the Green Toad frequents dunes, gravel pits, heaths, embankments and gardens. It breeds in still, sunlit waters, often using man-made ponds.

The call of the male carries far and can be heard mainly in April and May, most often at night. It is a shrill warbling interspersed with periods of silence and is reminiscent of the call of the Mole Cricket. Crepuscular and nocturnal outside the breeding season, the Green Toad leaves its daytime shelter under a stone or in a burrow to hunt its prey using agile jumps.

Common Tree Frog *Hyla arborea*

Anura
Length: 3 to 5 cm
Distribution: central and southern Europe.
Observation: March to October.

This frog is a good climber and favours habitats which have a combination of grass, bushes and trees. It breeds in still, sunny wetlands with plentiful vegetation, such as ponds, reedbeds, ditches and sometimes streams.

The male's powerful call carries a long distance, particularly when given as a chorus. It can be heard at nightfall from April to July, and is a halting croak reminiscent of a cicada. Outside the breeding season, the Common Tree Frog lives in bushy vegetation, most often between 40 cm and 2 m off the ground, and is active at dusk.

Marsh Frog *Rana ridibunda*

Anura
Length: up to 15 cm
Distribution: central and eastern Europe,
introduced into southern England.
Observation: March to October.

This frog lives in densely populated colonies and can be found in all freshwater habitats, from rivers and ditches to ponds and lakes. The call of the males can produce a deafening noise in big colonies – it is a typical loud and monotonous croaking sound, which can be heard predominantly from May to July. The Marsh Frog is principally diurnal, notably when hunting, but does sing at dusk. The Edible Frog (*Rana esculenta*) is a natural hybrid of the Pool Frog (*Rana lessonae*) and Marsh Frog – the first two are also common on the Continent and have small introduced populations in England.

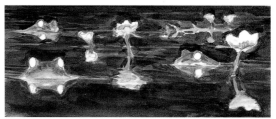

Stripeless Tree Frog *Hyla meridionalis*

Anura
Length: 4 to 5 cm
Distribution: south-west Europe.
Observation: February to December.

A sun worshipper, this frog colonises hot, sunny environments such as coastal marshes, humid scrub and sometimes parks and gardens. It breeds in reedbeds, ponds and sometimes streams.

The male's call is most commonly heard between April and July and is slower and deeper than that of the Common Tree Frog. It gives a loud, far-reaching *craaar*, which is often delivered in choruses at dusk and during the night. Less arboreal than its close relative, the Stripeless Tree Frog spends most of its time on the ground and is most active at night and in the morning.

Agile Frog *Rana dalmatina*

Anura
Length: 6 to 8 cm
Distribution: central and southern Europe.
Observation: February to October.

Although not always forest dwelling, the Agile Frog is strongly associated with broadleaved woodlands outside the breeding season. It breeds in a variety of different types of wetland, sometimes even settling for troughs or flooded meadows.

The male usually sings at night, especially in March, giving a rapid *car-car-car* that is difficult to hear when delivered underwater. Adults are mainly crepuscular and nocturnal and occupy a territory, moving around it in giant leaps and merging well with dead leaves.

Common Frog *Rana temporaria*

Anura
Length: 7 to 10 cm
Distribution: central and northern Europe, including Britain.
Observation: February to November.

A terrestrial species apart from during the breeding season, this highly adaptable frog can occupy numerous habitats from marshes and gardens to heaths and broadleaved and pine woods. It breeds in still or slow-moving water.

The male's call can be heard principally in March, sometimes in the day but mainly at night. It is a deep, monotonous, humming *grouk grouk grouk*. Outside the breeding season the Common Frog is active at day and during the night. It migrates en masse to its breeding sites on mild nights in late winter.

REPTILES

Moorish Gecko *Tarentola mauritanica*

Squamata
Length: 10 to 16 cm
Distribution: southern Europe.
Observation: almost all year.

This excellent climber is capable of hanging upside down thanks to the adhesive suckers on its feet. It frequents cliffs, rocks, woodpiles, dry stone walls and sometimes trees. Quite fearless, it often enters houses, where it is usually welcomed as a natural form of insect control.

It is active at dusk and during the night, although in autumn and winter it often basks in the sun during the day. It hunts insects such as mosquitoes and moths, and can often be found in the vicinity of streetlamps in towns and villages.

Slow-worm *Anguis fragilis*

Squamata
Length: up to 50 cm
Distribution: almost all of Britain and Europe.
Observation: March to October.

A legless lizard, the Slow-worm frequents open sunny habitats such as meadows, heaths, banks and gardens, as long as the ground is covered with dense herbaceous or bushy vegetation. During the day it hides under a stone, in a hole in the ground or even in a compost heap.

The Slow-worm is crepuscular but can be seen during the day, particularly in wet weather. It is rather retiring and moves slowly. It never bites to defend itself but will struggle vigorously and, like other lizards, can shed its tail. It remains active as long as the temperature stays above about 15°C.

Smooth Snake *Coronella austriaca*

Squamata
Length: up to 80 cm
Distribution: most of Europe. In Britain, very local in southern England.
Observation: March to October.

This species prefers dry, sunny, rocky habitats and can be found in woods, scrub, hedgerows, dry stone walls and vineyards. In England it occurs exclusively in a few areas of sandy heathland. Prey includes small mammals, lizards and snakes, including young vipers, which it detects by smell. Although the Smooth Snake can be seen hunting or sunning itself during the day, it is essentially active at dusk, most often during mild, wet spring weather, and during the night in hot weather. It is shy and apprehensive, moves slowly and is rarely aggressive.

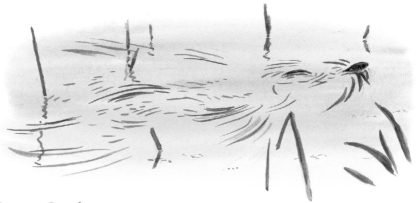

Grass Snake *Natrix natrix*

Squamata
Length: up to 1.20 m
Distribution: throughout almost all of Europe, including England and Wales.
Observation: April to November.

Although sometimes found in dry habitats, this snake is generally associated with wetlands and is often encountered near rivers, ponds, marshes and lakes. At home on land and in the water, it is a strong swimmer that feeds mainly on frogs and toads, although it also takes small mammals and fledglings, and even slugs. Principally diurnal, the Grass Snake occasionally adopts crepuscular habits, especially in hot weather and in the south of its territory. It is not aggressive but can produce a foul-smelling liquid to defend itself.

Aesculapian Snake *Elaphe longissima*

Squamata
Length: up to 2 m
Distribution: central and southern Europe.
Observation: March to October.

This snake frequents dry, sunny habitats, especially in the north of its range. Woodland, scrub, heaths, banks and old walls are all suitable. It hunts small birds, mammals and lizards, which it kills by constriction.

The Aesculapian Snake is principally diurnal but can be seen at dusk, especially in hot weather. Hunting on the ground, it also climbs up into bushes to search for prey and often raids birds' nests. It is usually calm and only becomes aggressive if threatened.

Adder *Vipera berus*

Squamata
Length: up to 80 cm
Distribution: much of Britain and central and eastern Europe.
Observation: February to November.

The Adder needs thick undergrowth and sunny areas for basking, which confines it to heaths,

hedgerows, scrub and similar habitats. It is shy and not very aggressive, only biting as a last resort if it cannot escape. Its venom is very rarely lethal to humans, but always requires medical attention. Most of the time this viper is diurnal, but in hot weather, especially in the south of its range, it can be active at dusk. It hunts on the ground, inspecting rodent burrows and other holes in search of the small mammals which make up the vast bulk of its diet.

Birds

It is easy to assume that most birds sleep at night, with the exception of the likes of owls and nightjars. In fact, though, a large number of species are active at night, especially during the breeding and migration seasons. It is also fascinating to find out where and how birds sleep. For example, the sight of a Penduline Tit waking up and leaving its reedbed nest at dawn, its feathers ruffled to keep out the cold, is an unforgettable one. Many birds sleep with their head under their wing, even larger species such as herons and raptors, while others like wildfowl and waders just slip their beak under the feathers on their back.

Dusk is an optimum period for viewing because so many birds are active and a simple piece of habitat such as a small wetland can be thronging with birds. Wagtails and buntings drop from the sky to roost in the reeds, while Moorhens and Water Rails call loudly. Gulls grouped on the water, preening themselves for the past couple of hours, suddenly taking flight towards a larger, safer lake for the night. Teal make the opposite journey to come and search for food, and snipe emerge from the reedbed and head for their feeding areas. Not to mention the Magpies sleeping in nearby trees, a Sparrowhawk making the most of the half light for one last hunt, and Blackbirds and thrushes making a terrific din as they come in to roost. Dawn is no less busy with the birds making the reverse journey, and to the accompaniment of a dawn chorus in the correct season.

During the migration periods, dusk often signals the departure of nocturnal migrants, which is three-quarters of all migratory species, while dawn heralds that of the diurnal migrants and the arrival of the last nocturnal ones.

Night Herons

Manx Shearwater *Puffinus puffinus*

Length: 32 cm
Distribution: Atlantic and North Sea. Breeds off west coast
of Britain and Ireland.
Observation: March to October.

The pelagic Manx Shearwater comes to land only
to breed in burrows on islets, returning at nightfall
to avoid predators and betraying their presence
with noisy, plaintive, hoarse calls. At dusk they
gather in rafts on the water just offshore. In the
Mediterranean, the Balearic, Yelkouan and Cory's
Shearwaters behave in a similar way, with the last
species giving a call that is reminiscent of a baby
crying.

European Storm-petrel *Hydrobates pelagicus*

Length: 15 cm
Distribution: Atlantic and Mediterranean coasts.
Observation: April to October.

Pelagic like the shearwaters, this storm-petrel is
even more elusive due to its small size. It returns
to its breeding island at night, nesting in a hole in
the ground or between rocks. Its characteristic
rolling call can be heard during courtship and
incubation. When breeding is over the birds return
to the high seas, feeding and sleeping on the water.
Winter is spent further south, at sea off the coast
of Africa.

Cormorant *Phalacrocorax carbo*

Length: 90 cm
Distribution: much of Britain and Europe.
Observation: all year. More widespread in winter.

Cormorants gather in evening roosts that sometimes contain several hundred birds. They are diurnal fish-eaters which spend the night in favoured trees that become whitened with their droppings. They will also roost on a bank, rock or island that is safe from predators. Their produce deep calls during the night as they quarrel amongst themselves for a prime perch. In the morning they disperse to their fishing grounds.

Greater Flamingo *Phoenicopterus ruber*

Length: 135 cm
Distribution: coasts of Iberia, southern France and Sardinia.
Observation: all year, although many birds winter in North Africa.

Even though its beautiful colours are not apparent in the dark, flamingos are often active at night. They do not require light in order to feed as they filter micro-organisms by touch, plunging their heads into the water of saltmarshes or rice fields. Its goose-like call confirms when it is active and it often flies at night. Flamingos are often nomadic and will cover long distances in order to find a suitable body of water.

HERONS

The different species occupy a variety of ecological niches in terms of fishing. Egrets are truly diurnal and form nocturnal roosts, while bitterns and Grey, Purple and Squacco Herons are more active as night falls and the Night Heron is the most nocturnal of all.

Night Heron *Nycticorax nycticorax*

Length: 62 cm
Distribution: southern and central Europe.
Very rare in Britain.
Observation: April to October, occasionally winters.

Its Latin name means 'crow of the night' and its large eyes are an indication of this species' nocturnal habits. After sheltering in the foliage of trees during the day the Night Heron flock leaves its roost as night falls and disperses towards its fishing grounds, with each bird punctuating its flight with frog-like *waak* calls. The species' favourite habitats are wooded marshes, reed-bordered canals, rivers and riverside vegetation.

Little Bittern *Ixobrychus minutus*

Length: 35 cm
Distribution: southern and central Europe – numbers fluctuate year to year. Very rare in Britain.
Observation: mid-April to September.

This tiny heron usually frequents reedbeds, but can be found on the banks of a canal or pond with reedy fringes mixed with willows. Its cry is a hoarse, muffled bark that sometimes sounds almost human. This is repeated every two or three seconds, can be delivered by day or night and is a sound that can easily merge in with the wider chorus of marsh creatures. Its nocturnal migratory call is slightly shorter and shriller than that of the Night Heron. From late August, birds prepare for migration by flocking. They become very vocal before flying up into the night sky and heading south.

Bittern *Botaurus stellaris*

Length: 65 cm
Distribution: locally across Europe. Scarce in Britain.
Observation: all year, but generally easier in winter.

The Bittern's call is a dull boom repeated six or seven times, which can be heard several km away

in calm weather. This characteristic reedbed sound can be imitated by blowing into an empty bottle. The Bittern is shy and only occasionally leaves the cover of the reeds, most often when forced to do so through frost or migration. When flying at night it gives a powerful *aaoh* cry that is reminiscent of a Great Black-backed Gull or Great Crested Grebe. The most likely time to catch a glimpse of it is at dusk, when it often goes fishing, appears on the edge of a reedbed or makes a short flight.

Grey Heron *Ardea cinerea*

Length: 95 cm
Distribution: throughout Britain and Europe.
Observation: all year.

The Grey Heron is commonly seen during the day, but at this time it is more often resting or preening than fishing. As dusk approaches, it often flies off to fish in a wetland such as a pond or river. It is more active at night when there is a bright moon or in places where there is street lighting. It has a keen eyesight and is difficult to approach – when disturbed it flies away squawking loudly and insistently.

WILDFOWL

Many of Europe's 40 or so species of ducks, geese and swans find food essentially using their sense of touch, and therefore do not require daylight in order to feed. This means that their daily routines are often determined by when their feeding grounds are least disturbed. Many species roost together in open water or the shelter of an island during the day, flying off at dusk to find food, sometimes providing a true avian spectacle in the process. Experienced at close range, the whistling of wings produced by a flock of ducks flying towards the fields, meadows or marshes is a memorable event. The calls of the birds add to the atmosphere and are a useful indicator of which species are present. Once they have landed, they dive, dabble or graze, but as they will be spooked if a torch beam is shone on them it is best to identify them

by their calls or by silhouette in the moonlight. Geese and some diving ducks, predominantly the fish-eating species, feed mainly during the day, and these are often joined by the species which are largely nocturnal feeders.

Mallard *Anas platyrhynchos*

Length: 55 cm
Distribution: throughout Britain and Europe.
Observation: all year.

The Mallard is the commonest European duck and is the ancestor of many domestic varieties. Its wings produce a characteristic whistling sound during flight. Only the female gives the familiar laughing *quack-quack-quack* call. The male emits a simple, dull *cwan* and whistling noises during courtship. It frequents all types of wetlands right down to the smallest pond or ditch, and can also be found in damp fields and sometimes crops.

Greylag Goose *Anser anser*

Length: 82 cm
Distribution: throughout Europe. Breeds in north and winters in south. Resident feral population in Britain.
Observation: all year.

Easier to spot during the day than at night, the Greylag Goose often flies at dusk between its feeding sites in meadows, crops and marshes and its roost sites on ponds, lakes and estuaries. Almost always in flocks, it can be identified by its loud call, which is similar to that of the domestic goose of which it is the ancestor. Birds roosting on the water tend to be very wary, but often betray their presence by flapping when preening and calling.

RAPTORS

As a general rule hawks, eagles, buzzards, harriers, kites, falcons and other raptor species hunt during the day and roost at night. Only the Peregrine Falcon has been observed hunting migrant birds at night under artificial lighting. Certain species such as hawks, harriers and falcons are crepuscular in their hunting habits, but they do not usually fly in the dark of night. Occasionally small species such as the Kestrel will migrate by night, particularly over the sea, but the larger species always roost, often returning each evening to a favourite perch. Kites, harriers and falcons often gather together in communal roosts.

Sparrowhawk *Accipiter nisus*

Length: 33 cm
Distribution: throughout Europe.
Observation: all year.

The Sparrowhawk skims hedges and borders at dusk and dawn, taking advantage of the smallest obstacle to hide its approach so that it can capture unsuspecting small passerines. It frequently hunts in towns and along river banks, spreading panic among roosting swallows, sparrows and starlings.

Hen Harrier *Circus cyaneus*

Length: 46 cm
Distribution: locally throughout Britain and Europe.
Observation: all year, but roosts communally during winter.

Hen Harriers disperse across fields, marshes and heaths to hunt rodents during the day, but flock together in the evening to roost, particularly in winter. As the sun sets they converge in a place with tall vegetation in which they can safely spend the night. So long as you remain in a secluded spot this is a good opportunity to see the birds well while they are preening or perched on a fencepost, and before they settle down. Montagu's and Marsh

Harriers also group together to roost, the latter in reedbeds.

Hobby *Falco subbuteo*

Length: 32 cm
Distribution: wooded countryside in most of Europe, including England and Wales.
Observation: April to October, winters in Africa.

Capturing dragonflies, swifts, swallows and martins during the day, the Hobby can also be found hunting bats at nightfall. Species which fly early and high, such as the Noctule, are the ones caught most frequently. When it nests close to the ocean the Hobby, like its relative the Eleonora's Falcon which breeds in the Mediterranean, scours coastal areas at dawn in search of migrant passerines that have not yet found shelter.

Osprey *Pandion haliaetus*

Length: 62 cm
Distribution: breeds in northern Europe, including locally in Britain. Also seen on passage en route to winter in Africa.
Observation: March to October.

Like many raptors, the Osprey often roosts in the top of a tall tree. Having no aerial predators to fear, apart from the Eagle Owl, an individual will return faithfully to the same perch time and time again if not disturbed. If approached with a good deal of caution they can be observed at quite close range in such favoured spots.

GALLIFORMES AND CRANE

Apart from the quail, which is active at night, most pheasants, partridges and grouse tend to be diurnal. However, all are early risers and their courtship and other activities frequently take place before daybreak.

Common Quail *Coturnix coturnix*

Length: 17 cm
Distribution: most of Europe, but scarce in Britain.
Observation: May to September.

It is difficult to pinpoint the source of the shrill, *wet-my-lips* song of the quail, but this and its weaker reedy call are characteristic summer sounds across much of Europe. The quail is active both day and night in crops and meadows where long grass keeps it hidden from view. It can also be heard calling in flight during the night, particularly in June.

Common Crane *Grus grus*

Length: 110 cm
Distribution: breeds in north-east Europe and winters in southern Europe, with a tiny resident population in Britain.
Observation: most spectacular at migration stop-offs and wintering sites.

Populations migrate south from October to December then return in February and March. Watching cranes arrive at and depart from their roosting sites is now a tourist attraction at various places in Europe, including the lakes in the Champagne region of France (a few thousand birds in winter and up to 25,000 in early spring), Matsalu Bay in Estonia (20,000 in autumn) and Gallocanta in Spain (50,000 in late winter).

After feeding in meadows and crops during the day, the cranes converge upon their roosting lakes from late afternoon until nightfall. Islands and mud banks provide them with safe places to rest and their bugling calls continue to ring out into the night, especially if there is a full moon or a flock of migrants comes in to land. At daybreak they take off and noisily disperse into the countryside.

Black Grouse *Tetrao tetrix*

Length: male 54 cm, female 42 cm
Distribution: peat bogs and heaths in the Alps and
northern Europe, including northern Britain.
Observation: all year, with lekking displays especially
from April to June.

In spring and summer the males arrive from all around
at first light to display at a communal lek. At dawn
they squabble and challenge each other, jumping up
and down to display the white undersides of their
tails, with excitement reaching fever pitch if a hen
appears. To see this spectacle it is best to locate a lek
that can be viewed from a car. Arrive during the night
and find a spot at a safe distance from the lek so that
you will not disturb the birds but can watch them
through a telescope. Do not leave the car as you will
disturb the lek. The show comes to an end soon after
the sun rises, but sometimes the birds also perform
at dusk.

Capercaillie *Tetrao urogallus*

Length: male 82 cm, female 60 cm
Distribution: high mountains in southern and central Europe,
more widespread in northern Europe. In Britain it is localized,
scarce and declining in the Scottish Highlands.
Observation: all year, with leks in April and May.

Normally solitary, these elusive giants gather each
year in spring to lek for a few weeks in the middle
of a pine forest. At nightfall the cocks fly in or arrive
on foot, sometimes singing, and perch in the trees to
roost. Once the sky begins to get light, they launch
into their hiccupping song, then alight on the ground.
From time to time they jump in the air, loudly beating
their wings to create visual (white underwings) and
sound (deep wing-beating) displays which are perhaps
more powerful than the song itself. These birds have
become extremely scarce in many places, and there
is now a code of conduct for watching them in
Scotland. The best way to see them is to attend an
event arranged at Loch Garten RSPB reserve or to
go on a specialist tour organized by a reputable
Scottish wildlife holiday company.

Marshland is one of the first habitats to come alive with birdsong in spring. A chorus emanates from the wintry reedbeds when the moon rises in the cooled air. Rails, crakes, Great Crested and Little Grebes, Moorhens and Coots are not really nocturnal: they do not actively search for food at night, but rarely miss a chance to make their presence known by calling or singing.

Corncrake *Crex crex*

Length: 23 cm
Distribution: much declined but breeds locally in central and northern Europe (in Britain mainly on the Hebrides), winters in Africa.
Observation: May to September.

The Corncrake population has been hit hard by modern farming methods and the species has disappeared from many of its former haunts. It shares similar habits, habitats and migration patterns to the Common Quail and, like that species, it is a difficult bird to see. In the event of danger it prefers to run for cover rather than fly, and often the only sign of its presence is the monotonous grating *crex-crex* call which can be heard especially at night. It frequents damp meadows and marshes.

Water Rail *Rallus aquaticus*

Length: 25 cm
Distribution: most of Britain and Europe.
Observation: all year.

It would be difficult to detect the Water Rail in its reedbed and marshland habitat if it wasn't so vocal. It proclaims its territory with a piglet-like squeal and gives a *kik kuk* alarm call when worried, while the meaning of the *kik kik kruuu* call drummed out endlessly throughout the night is less obvious. Quieter during the day, it becomes active at dusk and the most likely place to encounter it is at the margins of reedbeds. It sometimes calls while flying at night.

Little Crake *Porzana parva*

Length: 18 cm
Distribution: central and southern Europe, more common in the east. Vagrant to Britain.
Observation: April to September.

Male Little Crake (left) and adult Spotted Crake

Like all crakes, this species keeps a very low profile. The size of a Skylark, it never leaves the dense cover of the reedbed and sings most frequently at night. It can be heard most often between midnight and 2 am and its singing period lasts only a few weeks during spring, after which it becomes practically silent again. Its song (*whet whet,* followed by a descending series of similar notes) carries quite far, like that of the Spotted Crake (which gives a repeated *whit*!), while the croaking song of the Baillon's Crake is quieter and more easily confused with that of a Garganey or frog.

Moorhen *Gallinula chloropus*

Length: 30 cm
Distribution: all of Britain and Europe, except the far north.
Observation: throughout the year.

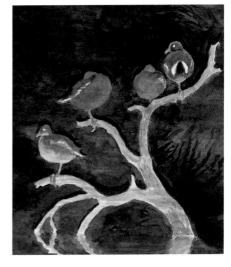

Active during the day and on clear nights, the Moorhen inhabits most types of wetland, including ponds, streams and rivers. Its trilled cackling call is loud and given frequently, especially when the birds are gathered in large numbers. A particularly impressive chorus occurs before daybreak in extensive marshes, where countless Moorhens call and reply to each other.

The Moorhen's white-billed relative, the Coot (*Fulica atra*), is almost as widespread and even more noisy. It cackles night and day as it displays, fights and patters across the surface of the water. Both species fly at night, calling regularly, migrating and colonising new wetlands.

WADERS

About 40 species of waders are regularly seen in Britain, although many of these breed much further north on the arctic tundra and only visit on passage or to spend the winter on our coasts. Different species have beaks of all shapes and sizes. Many feed by probing the silt, finding prey by touch, which enables them to feed during the day and at night. When flocks gather in bays and estuaries, they follow the pattern of the tide by feeding on the mud exposed at low water and sleeping when the water is high.

Certain species feed by sight and are therefore more diurnal than nocturnal, while others become more diurnal when nesting (Avocet and Oystercatcher), although others can be crepuscular or even nocturnal during breeding (Common Sandpiper and Curlew, while the Lapwing's display flight is often performed at night). Some species, such as Woodcock, Common Snipe and Stone-curlew, are primarily nocturnal.

Stone-curlew *Burhinus oedicnemus*

Length: 42 cm
Distribution: southern and central Europe. In Britain, scarce and highly localized in southern England and East Anglia.
Observation: March to October. Winters in Spain and Africa.

The Stone-curlew becomes most active at sunset. Its loud, shrill, rolling *cur-lee* cries are often the first sign of its presence in crops, lowland heaths, open stony wastes and areas of steppe. It has a parading posture, bowing its head and dipping its wings and tail, but it can be difficult to locate by day due to its excellent camouflage. It is much less active during the day, often settling down between two hummocks of earth, with which it blends in perfectly. It is best observed from a distance (to avoid disturbance) through a telescope at dawn or dusk.

Woodcock *Scolopax rusticola*

Length: 35 cm
Distribution: almost all of Britain and Europe.
Observation: all year, roding from March to July.

The *hon-hon-hon sip* call of the Woodcock heralds the return of spring to forest. Often merging in with the last calls of Song Thrushes, this sound could easily go unnoticed if it wasn't for the sight of the slow-flapping bird flying above the trees. This display flight, called roding, is carried out by males flying over part of the forest in search of females and it lasts well into the night and resumes at dawn. The Woodcock is diurnal and crepuscular when living in thick forest, and truly nocturnal when feeding meadows, heaths or other open habitats.

Common Snipe *Gallinago gallinago*

Length: 25 cm
Distribution: all of Britain and Europe.
Observation: all year, drums in May and June.

The Common Snipe is more nocturnal than is often realised. Many people assume that it is diurnal because it can be seen probing for food in broad daylight, especially in autumn when water levels are low. However, most snipe spend the day hiding in reeds or other tall vegetation, or in more open areas where they keep their heads down and stay well camouflaged thanks to their cryptic plumage. They become more active at dusk, calling to each other before disappearing into the darkening sky. They feed in neighbouring meadows or crops and return to the wetland before dawn. In the marshes where it nests, the snipe readily sings at night and executes a remarkable display flight as soon as the sky begins to darken. This is known as drumming and consists of alternating dips and curves, which are performed with the bird's tail completely spread open, and this causes the two outer feathers to produce a subdued bleating sound.

Black-headed Gull *Larus ridibundus*

Length: 40 cm
Distribution: all of Britain and Europe.
Observation: all year.

The Black-headed Gull is not a marine species, despite the fact that it is often seen on beaches and

near the coast. It breeds in colonies on lakes and marshes from March to August, and during this time it seems as if the birds never sleep because there is a constant barrage of noise, day and night. Outside the breeding season, Black-headed Gulls form night-time roosts on lakes, roof-tops or beaches. They often gather beforehand at a separate pre-roost site where they preen or bathe prior to continuing onwards to the place where they will spend the night. The light from streetlamps sometimes enables them to be seen crossing the night sky in towns.

Herring Gull *Larus argentatus*

Length: 60 cm
Distribution: Atlantic and North Sea coasts in Britain and northern Europe (and frequently seen inland in Britain).
Observation: all year.

Like the Black-headed Gull, Herring Gulls gather in large roosts and take advantage of artificial lighting to feed on rubbish tips or in ports, where they sometimes follow a boat out to sea. They utter a familiar loud laughing cry day and night at breeding sites, which are on marshes, cliffs, islands and the rooftops of coastal towns. At the last they can cause great annoyance to light-sleeping human inhabitants. A closely related species, the Yellow-legged Gull, breeds mainly around the coast of Iberia and the Mediterranean.

Feral pigeon *Columba livia*

Length: 32 cm
Distribution: towns throughout Europe.
Observation: all year.

The feral pigeon is one of the easiest birds to observe sleeping. It can be seen on many buildings at night with the help of street lighting. Other related species, such as the Collared Dove or Woodpigeon, can be found roosting in trees during the winter. It is often the latter which makes you jump in the forest at night, when you turn on a torch and it takes off and crashes through the branches with a loud flapping noise.

European Nightjar *Caprimulgus europaeus*

Length: 26 cm
Distribution: throughout Britain and Europe, except in the far north.
Observation: April to September. Winters in Africa.

The peculiar drawn-out churring of the nightjar can be heard at nightfall from May until August. The song is usually delivered from a stump or branch. The display also includes wing-clapping, and sometimes curious birds will come and fly silently around an observer. The flight action alternates rigid gliding beats with sharp turns for catching insects such as moths. Nightjars are fond of semi-open habitats with heather, including heath, scrub, forest clearings and open woodlands with pine or birch. At night birds that appear in a car's headlight beam show a bright eye-reflection long before the bird's silhouette can be distinguished.

OWLS

Owls have many adaptations to nocturnal life: large eyes to aid sight on the darkest of nights, facial discs to focus sound and soft plumage that does not make any noise during flight are all designed to enable them to capture hidden prey with precision.

Little Owl *Athene noctua*

Length: 22 cm
Distribution: central and southern Europe.
Observation: all year.

The most diurnal of its family, the Little Owl often perches in a sunny spot during the daytime, usually on the building or old tree that houses its roost. The species likes open habitat and hunts on the ground or perched low down. It can be seen in orchards, steppe and farmland bordered with walls or hedges. Its call is soft and melancholy, rising at the end with a final *quowic*, and its flight is undulating like that of the Green Woodpecker.

Barn Owl *Tyto alba*

Length: 36 cm
Distribution: all of Britain and Europe, except northern Scotland and Scandinavia.
Observation: year-round.

The classic image of the Barn Owl is of a bird captured in a car's headlights while perched on a fencepost or quartering a roadside verge. Its attraction to such habitat unfortunately means that it has suffered a huge toll of casualties from road accidents. Barn Owls require a large cavity for breeding, and attics and old churches and farm buildings are particularly suitable. They usually leave their roost at nightfall but sometimes hunt during the day, particularly in late winter and early spring. The call is a rather shrill, long and rasping hiss. The white-breasted form predominates in southern and western Europe, including Britain, while the dark-breasted form is more usual in central and eastern Europe.

Tawny Owl *Strix aluco*

Length: 40 cm
Distribution: Britain and Europe (but not Ireland or the Arctic).
Observation: all year.

The Tawny Owl's hoot is a characteristic sound of the night. The male typically makes a long *whoooo* followed by a short *hou* and ending with a quivering *whoo*. The female replies with a short or repeated *keevit* or a more shrill version of the male's song. The Tawny Owl often perches in a tree a few metres above a forest road, and can sometimes be located by attentively looking out into the beam of the car's headlights. Sightings are easiest at dusk in spring before leaves appear on the trees, or on summer mornings when groups of passerines mob an exposed owl with a barrage of alarm calls.

Pygmy Owl *Glaucidium passerinum*

Length: 17 cm
Distribution: central and especially northern Europe.
Observation: all year.

The tiny Pygmy Owl hunts small birds. Imitating its voice will set off distress calls from tits, wrens and other species in the forests where it lives. Its classic call is a slightly plaintive, light whistling *too-too-too* which can carry quite long distances and can be given during the day or at night, but is most frequent in the evening, especially in spring and autumn. The Pygmy Owl is not at all shy and will usually tolerate an observer standing at the foot of its tree. The slightly larger Tengmalm's Owl (*Aegolius funereus*) frequents many of the same forests and often roosts or nests in old Black Woodpecker holes. It is less inclined to answer an imitation of its call (a soft, rising tremolo).

Short-eared Owl *Asio flammeus*

Length: 36 cm
Distribution: all of Britain and Europe – essentially breeds in the north and winters in the south.
Observation: all year.

The Short-eared Owl is sociable and nomadic. Feeding mainly on voles, it wanders until it finds a place where prey is abundant. It quarters rough ground, marshes, heaths and meadows with a supple, low flight that is similar to that of large raptor such as a buzzard. It readily flies during the day, but is often active at dusk, sometimes rising in spirals and hovering on the wind to hunt prey. The female and young make a hissing that is deeper than that of the Barn Owl, while the song is a series of quick, deep hoots. It also gives a wing-clapping display during courtship.

Scops Owl *Otus scops*

Length: 20 cm
Distribution: southern and central Europe.
Very rare vagrant to Britain.
Observation: April to September.
Sings from April until mid-July.

The repetitive *tiu-tiu-tiu* song of the starling-sized Scops Owl can be heard in villages across southern Europe on spring evenings. Its simple fluted note is reminiscent of the Common Midwife Toad, but more powerful. It nests in the cavity of a wall or plane tree. Sometimes it can be seen hunting insects which have been drawn to a streetlamp.

Eagle Owl *Bubo bubo*

Length: 60 to 70 cm, wingspan 140 to 170 cm
Distribution: patchily across Europe. A tiny population of unknown origin has gained a toehold in Britain in recent years.
Observation: all year.

Restricted to rocky cliffs or stone quarries, the Eagle Owl spends its day stationary, roosting in a crevice, quiet nook or tree. Shortly before nightfall, the male strikes up its deep call, a low but powerful *who-oo*. It is this moment which offers the best chance of catching a glimpse of its big, cat-like silhouette outlined on the sky, before it glides away towards its hunting grounds.

Long-eared Owl *Asio otus*

Length: 35 cm, wingspan 90 to 100 cm
Distribution: throughout Europe.
Observation: all year.

The Long-eared Owl frequents a wide variety of environments including meadows, crops and marshes, provided there are sufficient copses or hedges for it to roost in during the day. It often sleeps in the shelter of a conifer and starts to hunt when night has almost fallen. It can venture out slightly earlier when feeding its young, whose typical shrill 'squeaky gate' calls can be heard from a long distance. The male's call is more discreet *hoo*, with which the female sometimes duets with a similar but quieter response.

In a chance roadside encounter it is not particularly sensitive to car headlights and can be watched for a period on a fencepost if you remain at a distance. At such times its ear-tufts are generally flattened down.

SMALLER BIRDS

Common Swift *Apus apus*

Length: 16 cm
Distribution: throughout Europe.
Observation: April to August.

The scythe-winged swift draws attention to itself by screaming loudly as it skims over rooftops before nightfall. When the evening performance ends, the breeding adults dive into their nests and the others rise up into the sky to sleep on the wing. It is quite possible to watch them at night by observing a full moon with a telescope in May or June. These birds make use of thermal currents and the wind, alternating occasional flaps with slow glides. In the morning they fly down to hunt near the ground.

Woodlark *Lullula arborea*

Length: 15 cm
Distribution: locally in Europe except the far north. In Britain, scarce but increasing on heathland in southern England.
Observation: throughout the year, but more frequent from March to October.

Smaller and shorter-tailed than its commoner relative the Skylark, the Woodlark has one of the most beautiful songs of all birds. Fluted and melancholic, it consists of a succession of descending notes, slightly hesitant to begin with, which become progressively louder and stronger over the course of a few seconds. It prefers to sing when the light is poor and can be heard frequently when it is overcast or raining, or at dusk and often during the night.

Pied Wagtail *Motacilla alba*

Length: 18 cm
Distribution: all of Britain and Ireland. White Wagtail
subspecies in continental Europe.
Observation: throughout the year.

Pied Wagtails are diurnal insectivores which often
spend the night together in huge communal winter
roosts that can contain several hundred birds.
These gatherings often take place in reedbeds,
where the wagtails are joined by other species such
as buntings, but many are situated in clumps of
trees in town and city centres. Such roosts are a
good time to monitor the birds' populations and
also provide quite a spectacle. The Pied Wagtail
(subspecies *yarrellii*) is common in Britan and Ireland,
while the paler-backed White Wagtail (subspecies
alba) is found across continental Europe.

Wren *Troglodytes troglodytes*

Length: 9 cm
Distribution: all of Britain and Europe.
Observation: throughout the year.

The Wren is a sedentary bird in Britain, where it
proclaims its territory with an unbelievably loud
song throughout the year. On long, cold winter
nights several individuals can gather together to
roost in order to stay warm. They huddle together
for the night in a hole in a bank or a bridge. A
nestbox is another favoured place, and the record
count is 63 Wrens in one box.

Nightingale *Luscinia megarhynchos*

Length: 16 cm
Distribution: southern and central Europe, including southern England.
Observation: April to September. Sings from April to June.

The Nightingale sings a song of alternating fluted crescendos, trills and accentuated notes as soon as it arrives at its breeding sites in April, as long as the temperature is not too low. After a short pause just after dusk, males can sing almost constantly through the night, as well as for a large part of the day. Males arrive from their African wintering grounds before the females and their song is intended to attract a mate and keep away rivals. Nightingales inhabit areas of thick vegetation such as scrub, bramble patches, thickets and hedgerows and are difficult to see. A closely related species, the Thrush Nightingale, occurs in eastern Europe.

Blackbird *Turdus merula*

Length: 24 cm
Distribution: all of Europe.
Observation: throughout the year.

The shrill call of the Blackbird is an integral part of soundscape in gardens and forests. Every evening just before dark the Blackbirds voice their insistent alarm notes as if worried by the approach of the night. Song Thrushes and Robins join in with their own calls and contribute to what can be a prolonged din, especially in spring. The chorus of their more melodious song typically begins before dawn in this season, even starting in the middle of the night in areas with street lighting, particularly when the weather becomes mild at the end of winter.

Bluethroat *Luscinia svecica*

Length: 14 cm
Distribution: breeds locally mainly in central and northern
Europe. More widespread on passage but a very scarce
migrant in Britain, where it is seen mainly on coasts.
Observation: March to August, winters in Spain and
further south.

A close relative of the Nightingale, the less well-
known Bluethroat sings frequently at night. The
song period typically begins in the middle of the
night and continues into the morning. It is a
versatile, jagged song which is difficult to decipher
when it merges with the chorus of other species
in the marshes. During the daytime the Bluethroat
accompanies its song with a courtship flight, which
is performed with wings and tail outspread. Females
and immatures have brownish markings on
the throat.

Redwing *Turdus iliacus*

Length: 21 cm
Distribution: breeds in northern Europe, winters in
southern, central and western Europe, including Britain.
Observation: October to April in Britain.

For many naturalists October is synonymous with
the calls of Redwings in the night sky. On a cold
night, a delicate *ziiih*, often answered by another,
betrays the arrival of these highly migratory
thrushes. The moon or lights in towns sometimes
allow a glimpse of the source of the calls. Going
out to listen for half an hour or so on a calm, clear
night in October will offer a good chance of hearing
Redwings and it may be possible to hear the calls
of other nocturnal migrants passing overhead and
to gauge the scale of passage. A good knowledge
of bird calls is a distinct advantage when doing this.

Grasshopper Warbler
Locustella naevia

Length: 13 cm
Distribution: breeds in central and western Europe.
A localized breeder in Britain.
Observation: April to September.

The Grasshopper Warbler is one of those 'little brown jobs' whose identification may at first appear very difficult to the novice. Most of these warblers sing with earnest through the night when they arrive back from their African wintering grounds and during the incubation period. The Grasshopper Warbler frequents marshes, heaths and open forests, and its song sounds similar to the stridulation of an insect. It is shrill, monotonous and very different to that of a Reed or Sedge Warbler. The bird delivers its song from a low branch, turning its head while singing, which makes it difficult for an observer to pinpoint its position.

Long-tailed Tit *Aegithalos caudatus*

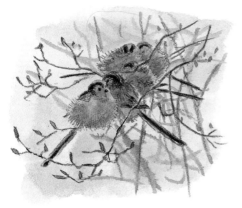

Length: 13 cm
Distribution: throughout Europe.
Observation: all year.

A cold winter's night is an ordeal for all small birds, but the highly sociable Long-tailed Tit has a way to combat this. Several individuals bunch together in a group at night in ivy or a dense bush, with feathers puffed out as much as possible and long tails protruding below. Other tits, Goldcrests and Nuthatches also group together to roost, often in a hole and particularly when the nights are cold. The different species have no qualms about mixing together.

Common Starling *Sturnus vulgaris*

Length: 21 cm
Distribution: throughout Europe.
Observation: all year. Gathers in large roosts especially from July to April.

Although not everyone appreciates the noisy, intrusive starling, it cannot be denied that watching massive flocks arriving at a roost in the evening is an impressive sight. Hundreds or thousands of birds whirl around in compact groups which bulge, stretch, thicken or separate like a coil of smoke before the birds suddenly come to rest in reedbeds or trees for the night. Then there is the hubbub of calls and songs mingling together, which can be heard on and off throughout the night. At dawn, the noise builds again, followed by a sudden silence that heralds departure, when thousands of wings beat and the sky blackens as the birds disperse in all directions.

Carrion Crow *Corvus corone*

Length: 47 cm
Distribution: western and central Europe, including Britain.
Observation: throughout the year.

The roosts of corvids, including Carrion Crows, Rooks, Jackdaws and Magpies, are often remarkable for the number of individuals that congregate. The birds often perch in the open and take flight at the slightest danger. Sometimes quiet, sometimes screeching, they fly around above an intruder until it goes away, whether it is a fox or a human observer. Several hundred birds whirling above you in the dark can create a striking impression and is not for the faint hearted! The Carrion Crow is replaced by the grey-and-black Hooded Crow (*Corvus cornix*) in eastern Europe, Ireland and western Scotland.

Mammals

Most European mammals are nocturnal or crepuscular. Apart from small mammals (small rodents and shrews), which feed during the day as well as at night, and a few diurnal species (such as squirrels, dormice and chamois), most mammals are only really active in low light conditions. Their most developed senses are those of smell and hearing, and this renders them more difficult to approach than birds, whose primary sense is sight (like humans). It is well worth sacrificing a little sleep in order to observe the complex and fascinating habits of mammals.

Mammal study methods are many and varied and include listening to calls with or without specialist equipment, observing and tracking them using tracks and signs, trapping and tagging, scanning with a light, and so on – many possibilities exist.

Some species can be watched by using a torch, although it is important not to disturb scarce or endangered species in this way. In general, animals with good eyesight are not particularly sensitive to artificial light. The advantage of this technique is that it enables the location of animals from a long distance thanks to their eye-reflection, which is caused by the tapetum – a layer of retina cells responsible for reflecting light. Once shining eyes have been detected, the torch can be aligned with a pair of binoculars for a better view of a small species or an animal at a distance. Certain birds, amphibians and even spiders can also be found using this method.

Recording mammal sounds is particularly useful as calls are difficult to describe and therefore also to memorise. This is also an activity in which an observer can spend long hours in the field without noticing the passing of time. Mammal sounds are so diverse that the recorded catalogue remains incomplete for many species.

Otters

Greater White-toothed Shrew *Crocidura russula*

Length: head and body 6-8 cm, tail 3-4 cm
Distribution: western Europe, but not Britain.
Observation: throughout the year.

Shrews live on the surface of the ground, remaining hidden under vegetation, stones or dead branches and avoiding direct sunlight. Their frantic pace and small size means that they must eat approximately every three hours. They can be heard at night, or spotted using a torch, but are easiest to observe during the day. The Greater White-toothed Shrew is one of the commonest European species and occasionally enters houses.

Water Shrew *Neomys fodiens*

Length: head and body 7-9 cm, tail 5-8 cm
Distribution: throughout almost all of Britain and Europe
Observation: all year.

The Water Shrew hunts by diving underwater in clear rivers, ponds or ditches. It is difficult to see as it shelters under roots and in bankside vegetation, becoming active mainly at dusk and during the night. On the other hand, its shrill, fitful calls are louder than those of other shrews and more easily betray its presence. The colouring of its underparts varies from white to dark grey. Underwater, the animal's fur collects a coating of air bubbles, giving it a gleaming appearance.

Pyrenean Desman *Galemys pyrenaicus*

Length: head and body 12-14 cm, tail 13-14 cm
Distribution: the Pyrenees and other mountain ranges
in northern Iberia.
Observation: throughout the year.

The desman is a primarily nocturnal inhabitant of mountain rivers and streams. It feeds on freshwater shrimps, caddisfly larvae and other aquatic invertebrates, which it captures underwater. Piles of droppings on a riverbank can indicate its presence, and early morning and evening are key times for observation. When searching it is best to find a place with calm water in order to more easily detect the ripples it makes when diving.

Hedgehog *Erinaceus europaeus*

Length: head and body 25 cm, tail 2-3 cm
Distribution: western and central Europe, including Britain.
Observation: March-April to November-December
(hibernates during winter).

An animal moving around noisily and unperturbed in the middle of the night is most likely to be a Hedgehog. Badgers or Wild Boar are sometimes just as noisy, but they rarely wait around to meet the observer. The Hedgehog comes out at dusk or after nightfall, roaming its territory in search of food. It is an animal with a remarkable biology and lifecycle, hibernating for part of the winter in cold parts of its range, protecting itself against predators by rolling up into a ball, and emitting gasping noises during territorial disputes. Finally, another strange habit is that it salivates abundantly while chewing on a range of objects, using this saliva to coat its spines. The closely related Eastern Hedgehog (*Erinaceus concolor*) is widespread in eastern Europe.

BATS

More than 30 species of bats can be found in Europe, with 17 of these occurring in Britain. All are insectivorous and hunt in flight at night. In order to do this, they have developed a complex echolocation system, making repeated high-pitched calls and waiting for the echoes to inform them about obstacles and prey. Some species fly before nightfall and are distinctive enough to be identified by sight. However, bats are most active during the dark, which means that specialist equipment is requiring in order to study them properly.

The best way is to use an ultrasound detector, which transforms the inaudible sounds into ones which can be perceived by our ears. It enables hunting bats to be heard and identified thanks to the frequency, rhythm and sonority of their calls. Another accessible but tricky method is to photograph the bats in flight using a small, compact digital camera. The camera must be preset to focus at two to three metres with a low zoom, and the flash must be activated. The bats can be seen approaching with the help of a low-power lamp, so as not to frighten them. Then it is a question of pressing the button at the right moment. Wooded borders, pathways and ponds are good places for watching bats. Certain caves or open buildings can serve as nocturnal resting places, especially on cool nights, or bats can also be seen in the evening leaving their diurnal roost, or at dawn whirling around in the sky before returning home. Remember that all bats and their roost sites are protected by law in Britain.

Common Pipistrelle *Pipistrellus pipistrellus*

Length: head and body 4-5 cm, tail 3 cm, wingspan 20 cm
Distribution: widespread in Europe.
Observation: throughout the year, but much less frequent between December and March.

By far the commonest of our bats, the Common Pipistrelle is a small species which can be seen flying close to houses in the evening or under streetlamps at night. During the day it shelters behind shutters or under the tiles of a roof. It comes out as night falls, but sometimes in the middle of the day in winter when the weather is mild. Several species of pipistrelle exist, including the Soprano Pipistrelle (*Pipistrellus pygmaeus*) which is also common in Britain. All are best separated with the help of a bat detector.

Brown Long-eared Bat
Plecotus auritus

Length: head and body 4-5 cm, tail 4-5 cm, wingspan 26 cm
Distribution: widespread in Britain and Europe.
Observation: April to October, hibernates in winter.

The Brown Long-eared Bat's large, forward-pointing ears can be easily seen in flight during a good view. The species leaves its roost in darkness soon after nightfall and returns home at dawn. Long-eared bats have an agile flight with gliding descents skimming the foliage, and they sometimes circle the same tree or building for several minutes. They are frequently found around flowering willows in spring and often hunt close to streetlamps. They alight under a porch or in an open building to rest or eat and dim lighting will not necessarily disturb them. The closely related Grey Long-eared Bat (*Plecotus austriacus*) has a more southerly range, and in Britain occurs only locally in southern England.

Greater Horseshoe Bat *Rhinolophus ferrumequinum*

Length: head and body 6-7 cm, tail 4 cm, wingspan 32 cm
Distribution: central and southern Europe. In Britain restricted to south-west England.
Observation: especially from April to October, hibernates in winter.

Despite its large size, the Greater Horseshoe Bat is difficult to observe away from its diurnal roost, generally emerging after nightfall and flying close to the ground until it reaches the cover of trees. It frequently uses a perch as a base for hunting, hanging from branches on the edge of a path or clearing. Greater Horseshoe Bats are extremely sensitive to disturbance in their hibernation or breeding sites. The are best watched at the rare sites where they are already familiar with the presence of people.

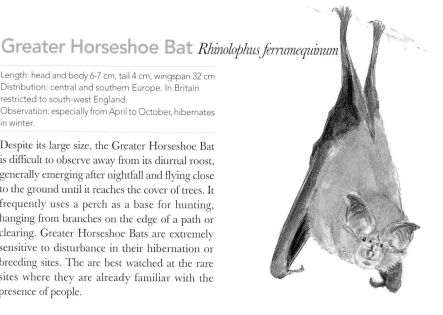

Noctule *Nyctalus noctula*

Length: head and body 6-8 cm, tail 5 cm,
wingspan 32-45 cm
Distribution: widespread in Europe, including
England and Wales.
Observation: especially from April to October,
hibernates in winter.

It is not uncommon to see Noctules flying in broad daylight, especially in autumn. They emerge early and often hunt over water, with their svelte silhouette and russet colouration both characteristic. The Leisler's Bat (*Nyctalus leisleri*) has a similar-shaped silhouette but is smaller and more localized in Britain (although it replaces the Noctule across Ireland). Both species commonly roost in a tree cavity in a forest or park, or sometimes in a building or bridge. The males call from their roosts to attract the females in early autumn, giving shrill, powerful calls which are discernible to many human ears. Likewise, summer colonies can be located during the day by their repetitive calls, with those of Leisler's Bat being more trilled.

Daubenton's Bat *Myotis daubentoni*

Length: head and body 5 cm, tail 4 cm, wingspan 23-27 cm
Distribution: most of Britain and Europe.
Observation: especially from April to October, hibernates
in winter.

A European bat seen skimming low over water in search of prey is most likely to be a Daubenton's. It flies close to the surface of ponds or rivers in order to catch insects, and sometimes picks them from the water. The pale belly of this species can be apparent if viewed with a torch and it is not particularly sensitive to the flash of a camera. Daubenton's Bats have been recorded living for up to 40 years.

Serotine Bat *Eptesicus serotinus*

Length: head and body 6-8 cm, tail 5 cm,
wingspan 32-38 cm
Distribution: much of Europe, including southern England.
Absent from the rest of Britain and most of Scandinavia.
Observation: especially from April to October, hibernates
in winter.

The Serotine Bat gives the impression of being
very large due to its big wingspan and slow flight.
Shortly before nightfall it leaves its roost, which is
most frequently located in the roof of a house. Its
silhouette is larger than that of the Noctule and its
flight more regular, with individuals often circling
in wide repeated loops. It regularly emits audible
calls, particularly when about to leave the roost,
and hunts in a range of habitats including villages,
open countryside or forests.

European Free-tailed Bat *Tadarida teniotis*

Length: head and body 8-9 cm, tail 5 cm,
wingspan 41 cm
Distribution: much of Iberia, plus regions bordering the
Mediterranean east to Greece.
Observation: possible throughout the year, but less
frequent from December to February.

The very large free-tailed bat is perhaps most easily
located by ear, especially by the powerful *tsic*
repeated about every second when an individual
has found prey. If it is still light, its narrow, angular
silhouette can be readily distinguished, with its
slender, arched wings enabling rapid flight. It roosts
in cliffs, buildings or bridges in which it occupies
fissures or uneven joints, but travels far to hunt,
sometimes even venturing out to sea.

Rabbit *Oryctolagus cuniculus*

Length: head and body 35-40 cm, tail 4-8 cm
Distribution: all of Britain and western Europe.
Observation: throughout the year.

Although the sight of a Rabbit in broad daylight is not unusual, the species is most active at dusk and during the night. Rabbits forage for food in the vegetation surrounding their warrens, and with each leap the white undersides of their tails glimmer in the dark. Their eyes reflect light beams very well and they are not usually spooked by artificial light. They tap the ground with their foot when worried or to signal danger. Depending on the density of the population, Rabbits live pairs or in colonies with a dominant male and female.

Brown Hare *Lepus europaeus*

Length: head and body 40-70 cm, tail 7-13 cm
Distribution: Britain and Europe, except the far north. Usually at lower elevations than Mountain Hare, although can occur quite high in mountains.
Observation: all year.

In the evening large fields can seem to come alive with hares, with animals appearing all over the place. The Brown Hare often remains motionless during the day, sheltering in a shallow scrape in the ground or simply lying down against a hillock. It becomes active as night approaches, but often earlier in winter, and frequents woods, fields or crops. When snow falls, a succession of characteristic Y-shaped tracks reveal the species' favourite pathways. Its eye gleams when lit up by a torch and its immense black-tipped ears make it easy to distinguish from the Rabbit. In the courtship season between late winter and spring, 'mad March hares' chase each other at length and have 'boxing' fights while standing up on their hind legs.

Mountain Hare *Lepus timidus*

Length: head and body 45-60 cm, tail 4-8 cm
Distribution: Alps and northern Europe, including Scotland, Ireland and northern England.
Observation: throughout the year.

Immediately identifiable in winter by its pure white coat (except in Ireland), the Mountain Hare is distinguishable from the Brown Hare in summer thanks to its large, light-coloured paws, smaller ears and small, entirely white tail. The Mountain Hare is a crepuscular and nocturnal animal; a mountain-dweller which is hard to observe. A hide set up on mountain pastures in spring can afford good views, on the condition that you spend the night at high altitude. In the far north, Mountain Hares live at lower altitudes and are necessarily diurnal during the long hours of summer daylight.

Porcupine *Hystrix cristatus*

Length: head and body 60-80 cm, tail 6-12 cm
Distribution: Italy.
Observation: throughout the year.

A native of Africa, the Porcupine was introduced into Europe by the Romans and still lives locally in Italy. This enormous vegetarian rodent lives underground in a burrow by day and wanders through fields, wilderness and woods at night. Its stiff, light-coloured quills give it an impressive appearance in the moonlight, which is reinforced by the rattling of its armour. It stands its ground and bristles its quills when frightened.

European Beaver *Castor fiber*

Length: head and body 75-90 cm, tail 30-40 cm
Distribution: scattered populations in eastern and
northern Europe. Small-scale reintroduction projects
in progress in Britain.
Observation: throughout the year.

The beaver is perfectly adapted to life in the water, but because of this it has become dependant on its aquatic environment. It rarely ventures further than a few metres from the water to cut down a tree or a branch. Willows are a favourite and it returns regularly to the shoots derived from the branches it cuts. The beaver pulls the cuttings into the water and eats them while half-submerged, ready to dive in the event of danger. It sometimes drags shoots through the water to a place where it feels safe. This feeding area can be spotted during the day thanks to the cut remains of the stalks, and it can be visited by all the family, including young from the present year and the previous year. Beavers shelter in a burrow dug in a bank, or in a lodge, which is a mound of floating branches that supports a nest built just above the water-level.

The animal emerges at night, letting itself drift along while smelling the air for potential danger on the bank, then pulls itself up onto dry land to eat and groom. The best sites for watching beavers are, paradoxically, those which are often visited by humans, where the creatures have become less wary and will swim only a few metres away from bathers. They tolerate torchlight and can also be seen in the morning before they dive off towards a daytime shelter.

Muskrat *Ondatra zibethicus*

Length: head and body 25-40 cm, tail 20-28 cm
Distribution: introduced in central and eastern Europe
(eliminated from Britain in the 1930s).
Observation: throughout the year.

Introduced from North America, the Muskrat is found in a variety of wetlands, including ponds and rivers. It comes out mainly at night, but can be active during the day. A lodge is constructed from reeds in places where it cannot dig a burrow. When it swims on the surface of the water, its tail trails along behind it. Like all aquatic rodents, it dives well and disappears at the first sign of danger.

Coypu *Myocastor coypu*

Length: head and body 40-60 cm, tail 30-45 cm
Distribution: introduced from South America. Found locally in continental Europe but eliminated from England in 1990s.
Observation: throughout the year.

More diurnal than the beaver, the adaptable Coypu is at home in almost all aquatic habitats and has colonised streams, ponds and fresh- and saltwater channels in various parts of Europe. Its gherkin-shaped droppings and well-trodden walkways through bankside vegetation are two indicators of its presence. Coypu can be distinguished from beavers when swimming because their backs protrude above the surface of the water and their heads are more angular. They also have white whiskers, which distinguish them from both beaver and the smaller Muskrat. Calls are frequent and include a sort of low mooing and wailing and a sound similar to a baby crying. Coypu shelter in burrows and are quite tolerant of torchlight.

Southern Water Vole *Arvicola sapidus*

Length: head and body 17-22 cm, tail 12 cm
Distribution: France and Spain. The shorter-tailed Northern Water Vole (*Arvicola terrestris*) is found in the rest of Europe, including Britain.
Observation: throughout the year.

This small aquatic rodent it is at home in streams and even shallow ditches as long as the vegetation on the banks provides adequate food. It is active at dusk and night, but also during the day, running and swimming along the banks, climbing to gather the plants it likes and diving into the water in the event of danger. Shrill cries like chicks cheeping are evidence of interaction between individuals.

Edible Dormouse *Glis glis*

Length: head and body 13-19 cm, tail 12-15 cm
Distribution: central and southern Europe, but rare or
absent in much of western Europe. Introduced and
spreading in southern England.
Observation: April-May to October, hibernates in winter.

The source of a shrill, grating squeal repeated at
length from high in the branches is likely to be this
dormouse. Other individuals often reply to a call,
and with a little luck you may be able to find one
in a torch beam without scaring it. The light
underparts show up well in torchlight, but the
animal can disappear in a few leaps through the
branches. The Edible Dormouse is nocturnal and
arboreal. It lives on wooded hillsides and rocky
cliffs, and often in buildings provided that there
are trees nearby, particularly beech.

Garden Dormouse *Eliomys quercinus*

Length: head and body 11-17 cm, tail 9-15 cm
Distribution: central and southern Europe.
Observation: April to October, hibernates in winter.

As vocal as its relative, the Edible Dormouse, this
species does not live in forests, but prefers hedged
farmland, orchards and villages, as well as rocky
habitats in the Alps. It often hibernates in houses.
It is very alert and difficult to observe. It can run
at high speed through dense vegetation and does
not usually tolerate torchlight. When threatened
it gives shrill calls that are more muffled than those
of the Edible Dormouse: an endlessly repeated
tchrrrjj every one to three seconds, or the same call
given in a shrill, prolonged trill.

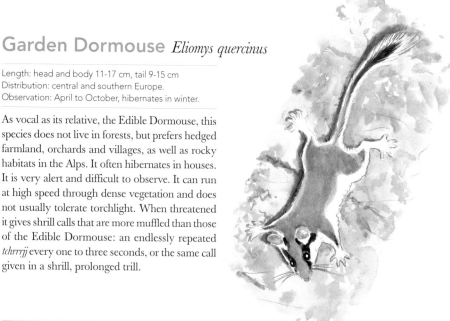

Common Dormouse *Muscardinus avelanarius*

Length: head and body 6-9 cm, tail 5-8 cm
Distribution: central and southern Europe, including
locally in southern Britain.
Observation: April-May to October, hibernates in winter.

This dwarf of the dormouse family is extremely shy
and elusive, and mostly silent. Although active at
night, there is a better chance of spotting it during
the day by closely inspecting brambles and clematis
in woodland clearings. It is here, in dense vegetation,
that the animal can be glimpsed slipping out of its
small, ball-shaped nest and remaining still while
waiting for the observer to leave. If you remain still
and a few metres away, the Common Dormouse will
return to its nest after a few minutes. It will leave
again at nightfall to go and feed on buds, berries and
insects. It frequents mixed forests, reedbeds and other
well-vegetated habitats where it can move about in
the branches without having to jump. Populations
are more dense at the end of summer.

Harvest Mouse *Micromys minutus*

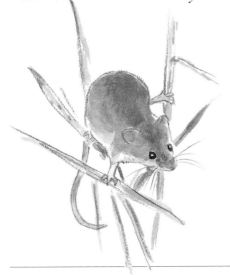

Length: head and body 5-8 cm, tail 5-7 cm
Distribution: much of Europe, including
England and Wales.
Observation: throughout the year.

A tiny rodent, the Harvest Mouse builds its nest
in tall tangled grass, reeds, brambles or hedgerows.
The size of a tennis ball, the nest incorporates green
leaves from surrounding plants which are shredded
lengthwise. The Harvest Mouse is active at dusk
and during the night, but its small size makes it
very difficult to see in the dark. Its call is a restrained
cheep, which is less loud and shrill than that of a
vole or shrew.

Common Hamster *Cricetus cricetus*

Length: head and body 20-34 cm, tail 2-3 cm
Distribution: central and eastern Europe.
Observation: March-April to October, hibernates in winter.

The hamster is a colourful and sizeable rodent which is nocturnal and not particularly easy to observe, although its burrows are a good clue to its presence. Its favoured habitats are farmland, steppe and other dry grassland and it can cause crop-damage, particularly to alfalfa and clover fields. Huge quantities of food can be stored underground after being carried there in its elastic cheek-pouches. It leaves its burrow at nightfall, but sometimes earlier, particularly in early spring. It is very tolerant of torchlight.

Bank Vole *Clethrionomys glareolus*

Length: head and body 8-12 cm, tail 4-7 cm
Distribution: widespread in Europe, including Britain, but not Spain and Greece.
Observation: throughout the year.

The Bank Vole is easily identified by its grey and rufous fur, and it lives in all sorts of forests from lowlands to mountains. It is easiest to see at the beginning of the spring when there is less leaf litter and when shorter grass and a scarcity of food means that it is more likely to venture out into the open. The Bank Vole comes out during the day as well as at night, and is often most active at dusk. It can climb several metres high in a tree in order to feed on bark.

Common Pine Vole *Pitymys subterraneus*

Length: head and body 8-10 cm, tail 3-4 cm
Distribution: central Europe.
Observation : all year.

A small-eyed, greyish rodent, the Common Pine
Vole is difficult to see as it is more nocturnal than
the Bank and Field Voles and rarely ventures out
into the open. It lives in grassy verges, borders and
forests, and especially favours damp conditions.

Common Vole *Microtus arvalis*

Length: head and body 9-12 cm, tail 3-4 cm
Distribution: much of Europe, but in Britain only on Orkney.
The similar Field Vole (*Microtus agrestis*) is widespread in
Britain and Europe.
Observation: throughout the year.

The Common Vole shares much of its range with
the closely related Field Vole. Both live in crops,
meadows, pastures and verges, although the former,
which is longer-tailed, favours habitats which are
rather more open and dry than those frequented
by the latter. Both species have several periods of
rest and activity during the day, but rarely venture
out in broad daylight. They are easiest to observe
during their periodic population explosions, which
attract good numbers of their predators such as
raptors, owls herons and foxes.

Wood Mouse *Apodemus sylvaticus*

Length: head and body 8-10 cm, tail 7-11 cm
Distribution: all of Britain and Europe, except the far north.
Observation: throughout the year.

Although present in a wide range of habitats, the Wood Mouse often goes unnoticed as it is essentially nocturnal. It lives in forests, gardens, heaths and dunes, and can be found in mountains up to altitudes of 2,500 m. It is particularly active around dusk and at the beginning of the night, but less so on cold, moonlit winter nights. During the night it can often be heard running around in dead leaves even though it can't be seen. The Wood Mouse can occasionally be spotted crossing the road in the beam of a car's headlights. Occasionally it gives a shrill, short call, and it is tolerant of torchlight. It often enters houses in places where it replaces the House Mouse, and it can be distinguished from that species by its larger eyes and ears.

House Mouse *Mus musculus*

Length: head and body 7-10 cm, tail 7-9 cm
Distribution: throughout Europe, with subspecies *domesticus* in the west (including Britain) and subspecies *musculus* in the east.
Observation: throughout the year.

The House Mouse is not particularly common in modern houses, where it can be confused with the sporadic visits of field mice and shrews. Once settled its presence is betrayed by its droppings and odour, and by the damage it does, notably to foodstuffs. It comes out mostly at night and avoids brightly lit areas. Once it becomes familiar with the regular noises in a house it can sometimes lose much of its wariness.

Brown Rat *Rattus norvegicus*

Length: head and body 21-29 cm, tail 17-23 cm
Distribution: throughout Britain and Europe.
Observation: all year.

The Brown Rat is the biggest of Europe's mice and rats, and is also the easiest to see. It often lives in close proximity to humans, exploiting food from farming, rubbish dumps and food storage sites, but it also lives in cellars and drains and on the banks of canals, rivers or ponds. It climbs and swims well, and in the water its long silhouette and pointed head distinguish it from the water voles or Muskrat. It generally comes out at night, depending upon the availability of food, the presence of predators (such as cats, foxes or Beech Marten), and the hierarchy within the group if the population is big. Street lighting often provides good opportunity for observation.

Black Rat *Rattus rattus*

Length: head and body 16-24 cm, tail 12-26 cm
Distribution: much of Europe, but now very rare and localized in Britain.
Observation: throughout the year.

Slimmer and with bigger ears, the Black Rat is less well adapted to the cold than its commoner relative. Away from the Mediterranean countries it is most commonly found around towns and ports. It is nocturnal and a good climber, spending a good deal of time feeding on fruit and seeds in trees in hot climates. In houses, it prefers the upper storeys and often explores roofs and attics, while it also visits storerooms to take advantage of domestic foodstuffs. Like the Brown Rat, the Black Rat can perceive and emit ultrasound.

Wolf *Canis lupus*

Length: head and body 90-150 cm, tail 30-50 cm
Distribution: northern and eastern Europe, with isolated populations in Spain, Italy and France.
Observation: all year.

As a competing predator at the top of the food chain, the Wolf has traditionally invoked fear and respect in equal measures from humans, and also suffered persecution to the point of extinction in many parts of its former range, including Britain. After centuries of being a constant victim to hunters it rarely allows itself to be seen. The Wolf has survived through adapting its behaviour and becoming more nocturnal where it continues to live close to humans. It covers miles of territory on its hunting trips and sometimes even scavenges at rubbish tips or places where food has been put out for bears. Often the first sign of a Wolf is its tracks, droppings or calls. They howl at night, either alone or in chorus, and most often in winter.

Red Fox *Vulpes vulpes*

Length: head and body 60-80 cm, tail 40-50 cm
Distribution: throughout Europe.
Observation: all year.

The fox lives in a wide variety of habitats, from coastal marshes and city centres to lowland forests and mountain meadows. It adapts its diet according to local conditions and maintains its population even in places where it is hunted intensively by humans. Although nocturnal by necessity in many places, it often ventures out in the evening and hunts well into the morning if undisturbed, foraging in dustbins, freshly mown fields where voles are vulnerable, and on lawns for earthworms in parks and gardens. It is very active in the January and February mating season despite the cold. Perhaps the most common call of its varied vocabulary is the vixen's shrill and powerful *whaaa*, which is repeated every few seconds. The male gives a quieter *wo wo wo*, especially in the breeding season. Fox cubs leave the den during the day when they are very young (at the beginning of May), and at nightfall when older. Like most carnivores the fox is not sensitive to torchlight and its eyes reflect the light from a distance.

Brown Bear *Ursus arctos*

Length: head and body 170-250 cm, tail 10-20 cm
Distribution: scattered populations in southern,
eastern and northern Europe.
Observation: all year, but hibernates locally from
November-December to March-April.

Just knowing that bears are present adds an extra dimension to wildlife-watching. Coming across the enormous footprints is impressive enough, but actually seeing the animal itself is something else. Getting a view is very difficult without using bait, although hides with food such as corn or carrion are being set up to watch or photograph bears in increasing numbers of countries where they occur.

Bears are essentially nocturnal and have remarkable sight and hearing. They actively avoid humans, apart from a few curiously indifferent individuals, and are very sensitive to artificial light. It is an omnivorous forest animal, most often occurring in mountains, and only ventures out into the open at certain periods to search for food such as roots and berries.

Pine Marten *Martes martes*

Length: head and body 35-55 cm, tail 18-28 cm
Distribution: widespread in Europe, but absent from large parts of southern Britain and Spain.
Observation: throughout the year.

The Pine Marten is the most arboreal of European mustelids and rarely ventures out into the open. Never abundant, it crosses its large territory with characteristic bounds and leaps. It is active well before nightfall and after sunrise in spring and at the beginning of summer, but is more nocturnal for the rest of the year.

A diet of rodents, birds and invertebrates is supplemented by fruit and berries in summer and autumn. It shelters in a tree hollow or abandoned nest, and sometimes moves from tree to tree like a squirrel. Forests on flat ground with long, straight paths are good places for observing Pine Martens, particularly at dusk.

Western Polecat *Mustela putorius*

Length: head and body 30-45 cm, tail 13 cm
Distribution: widespread in Europe, but absent from the arctic, Ireland and parts of southern Britain.
Observation: throughout the year.

An elongated body and dark face-mask make the polecat easily identifiable. It is crepuscular and nocturnal but can be active during the daytime, especially in forests. Forest tracks and paths around ponds or marshes are the best places for catching a glimpse of one. It is not sensitive to torchlight and can come right up to the observer. Its sense of smell and hearing are sharper than its eyesight and these enable it to locate the rodents and amphibians that make up the bulk of its diet. There is a clear size difference between the sexes, with males being larger.

Beech Marten *Martes foina*

Length: head and body 45 cm, tail 26 cm
Distribution: most of Europe, but absent from Britain.
Observation: throughout the year.

The Beech Marten can be distinguished from the Pine Marten by its white bib, smaller ears and shorter legs. In general it has a thinner and lighter-coloured snout, with a pinkish nose instead of a black one. As good a climber as its relative, it often shelters in attics and roof cavities in houses, whether they are inhabited or not.

It is present in woodlands, rocky hillsides and towns and villages and its behaviour is particularly nocturnal at sites in which it lives in close proximity to humans, where it comes out as night falls and returns to its den before daybreak. The best chance of spotting one is therefore with a torch, and in many places it has become accustomed to artificial light thanks to street lighting. Males are especially active marking their territory between May and July, increasing the chances of an encounter during this period.

American Mink *Mustela vison*

Length: head and body 30-45 cm, tail 15-23 cm
Distribution: American introduction to Europe from Spain to Scandinavia, including Britain.
Observation: throughout the year.

The American Mink, which frequently escapes from captivity while being bred for its fur, has become widespread in many countries in contrast to the European Mink (*Mustela lutreola*) which has become scarce in many places and extinct in much of western Europe. Mink are closely linked with wetland habitats such as marshes, rivers and coasts. They are excellent swimmers, diving and catching fish and shellfish. They also explore banks and rocks to hunt small animals and have been implicated in the decline of the Northern Water Vole in Britain. They are predominantly crepuscular and nocturnal.

Lynx *Lynx lynx*

Length: head and body 80-130 cm, tail 12-25 cm
Distribution: locally in northern, eastern and central Europe.
The smaller Iberian Lynx (*Lynx pardina*) is rare in Spain.
Observation: throughout the year.

A shy but not particularly aggressive forest hunter, the Lynx is crepuscular and nocturnal and feeds mainly on large prey such as Roe Deer, which it consumes over a period of several days. It sleeps during the day, concealed among rocks or in thick bushes, and comes out at nightfall to return to its previously killed prey or to search for a new victim. It is capable of covering considerable distances in search of food or a mate. A male's territory can span 400 sq km, and during the courtship period in February and March they roam huge distances calling regularly. The call is a shrill *waaoh*, reminiscent of a vixen but more guttural and less violent. Young Lynx stay with their mother for about one year.

Stoat *Mustela erminea*

Length: head and body 20-30 cm, tail 10-14 cm
Distribution: Europe, except Mediterranean regions and southern Europe.
Observation: throughout the year.

The Stoat, like the Weasel (*Mustela nivalis*), is active both day and night, alternating short periods of activity with long rests. However, given the small size and sprightliness of these two species, there is a much better chance of spotting them during the day than at night.

Wildcat *Felis sylvestris*

Length: head and body 50-70 cm, tail 20-40 cm
Distribution: Scotland, eastern France and western
Germany and across southern Europe.
Observation: throughout the year.

Distinguishing the Wildcat from the domestic cat can be quite difficult, but the Wildcat has longer legs, longer ears, a bigger body and a thick, ringed tail. It hunts in forests and wooded borders where it feeds mainly on voles. It is most likely to be seen out in the open during periods of snow or shortly after fields have been mown or crops cut. Otherwise, the best chance of spotting it is from a well-situated hide or on a forest pathway, as in the undergrowth it generally locates the observer before it is seen. The Wildcat is slightly more confident in the dark and is not sensitive to torchlight.

Common Genet *Genetta genetta*

Length: head and body 45-60 cm, tail 40-50 cm
Distribution: Iberia and western France.
Observation: throughout the year.

The agile genet can hunt in trees and on steep cliffs. It can sometimes be found in ditches or hedgerows, but its favourite hunting grounds are ravines and woodlands. Its large droppings typically have grass at either end and are deposited in elevated latrines, which are usually situated on a large rock or a tree stump. The genet is principally nocturnal but also crepuscular and is most often seen in the beam of a car's headlights. However, it is also often possible to get views by walking quietly with a torch.

Female and young Badger encounter a fox

Badger *Meles meles*

Length: head and body 65-80 cm, tail 11-19 cm
Distribution: all of Britain and Europe, except the far north.
Observation: throughout the year. Slightly less active in
December and January.

The Badger is at its most impressive when lit up by the moonlight. Its black-and-white head gleams in the dark and its grey coat is fringed with light and dark reflections due to the effect of the light on its long hairs. In better light, however, its most noticeable features are its small, pig-like eyes, thick nose and dirty white snout. Watching a Badger sett requires quietness and patience, as well as locating yourself with the wind in the right direction so that the animals cannot detect your presence. After hesitating for a few moments, the Badger slides out of its hole and generally begins grooming itself. When a second appears they sometimes tussle with one another before going their separate ways. They can then be seen going about their business in fields, undergrowth, woods or thickets, where they feed on plants, earthworms and other invertebrates. Badgers are nocturnal and only come out at dusk, although during spring and early summer times varying greatly according to how undisturbed the area is. They return to their sett during the night or at dawn. By observing a sett it may be possible to see them grooming or removing unwanted debris. Badger cubs first venture above ground around mid-April and accompany their mother for several months. Badgers are generally rather sensitive to torchlight.

Otter *Lutra lutra*

Length: head and body 60-90 cm, tail 35-45 cm
Distribution: widespread in Britain and Europe, but absent from many places.
Observation: throughout the year.

The Otter is a carnivore which has adapted to aquatic life and feeds on fish and shellfish. It has a vast hunting territory and occupies all wetland habitats from streams to coasts. It covers several kilometres each night by swimming and running, and marks its territory with small droppings known as spraints as it goes. The Otter is crepuscular and nocturnal, although there is a great deal of local variation and some individuals are more diurnal.

The young accompany the female for almost a year. Contact calls are frequent, the most common being a shrill whistling that is repeated regularly about every two seconds. Audible at a distance of 200 m in calm weather, its intensity diminishes rapidly if the animal turns its head or moves behind an obstacle. The Otter swims quickly through the water with its head and the base of its back remaining slightly above the surface, and it dives silently. It can, however, be very noisy when foraging on the shore or playing. The best way to view Otters is by moonlight along one of their regular runs, but the chosen spot should be where the water is not too noisy (to give a better chance of detecting the animals) and with the wind in the right direction.

Otter scent-marking

With several hundred species of marine fish and invertebrates, the ocean is a separate world that is too vast to cover in this book. Species occupy different niches according to depth, current and the material on the seabed, and observation of them requires specialist knowledge and equipment for navigating and diving. However, there are some larger creatures which can be seen on the surface. Seabirds such as storm-petrels and shearwaters were covered in the previous section on birds, while a small selection of the larger sea mammals is shown here.

Striped Dolphin *Stenella coeruleoalba*

Length: 2-2.5 m
Distribution: Atlantic and Mediterranean.
Observation: throughout the year.

Dolphins swim alongside boats at night as well as during the day. Light from the deck can illuminate them bow-riding in the black waters, and they can sometimes be heard blowing, jumping and calling.

This species is common well offshore, with the plainer grey Bottle-nosed Dolphin and the diminutive Harbour Porpoise much more likely to be sighted from land.

Grey Seal *Halichoerus grypus*

Length: 2.1-3.3 m
Distribution: coasts of Britain and Europe from north-west France to Norway.
Observation: all year.

Grey Seals are active at night and rest during the day, especially in late afternoon, although during the autumn and winter breeding season they are usually active day and night. Females gather together on a sandy or rocky shore and the males endeavour to defend small groups of them from rivals. Grey Seals give long howls and growling calls, especially during the breeding season.

Fin Whale *Balaenoptera physalus*

Length: 18-23 m
Distribution: Atlantic and Mediterranean.
Observation: all year.

The sight and sound of a whale-blow at sea on a calm, starry night is a magical and inspiring one. The Fin Whale, like all large cetaceans, dives to great depths to feed and expels the air from its lungs when it surfaces. This creates an impressive blowing sound, which can be heard from a great distance on a calm sea.

Wild Boar *Sus scrofa*

Length: head and body 100-160 cm, tail 15-30 cm
Distribution: Europe except Britain, Ireland and
Scandinavia, although there are small feral populations
in England.
Observation: all year.

Scuffling in the undergrowth, branches cracking and grunting noises indicate the presence of a group of boar searching for food. They are not fussy eaters and will take bulbs, acorns, chestnuts, corn, earthworms and various small animals. Boar predominantly live and sleep in areas of dense vegetation such as forests and marshes with reedbeds, but they often venture further afield at night to feed in fields and crops. However, they always return to a quiet, sheltered spot by daybreak.

Groups are made up of females and young, with males gradually excluded to become solitary with age. They will not charge humans, even when accompanied by young, unless seriously threatened. A disturbed boar will snort noisily and loud grunt will often be the signal for a group to flee. Wild Boar frequently fight to the accompaniment of aggressive calls, and they are not sensitive to torchlight.

Red Deer *Cervus elaphus*

Length: head and body 160-240 cm, tail 15-20 cm
Distribution: scattered populations across Britain
and Europe.
Observation: throughout the year.

More crepuscular than nocturnal, both stags and hinds often leave the forest at dusk and alternate periods of grazing and rest before returning to the safety of cover by dawn. The herds are made up of adult and immature hinds, year-old calves and fawns of the present year. The stags live separately, either alone or in groups, and join the herds towards the end of August when the rut begins.

At this time the stags roar loudly and fight using their antlers in order to try and keep as many hinds as possible close to them for mating while simultaneously driving their rivals away. This exhausting event continues until late October or November when the forest again falls silent, disturbed only by the tremulous hoots of Tawny Owls.

Roe Deer *Capreolus capreolus*

Length: head and body 95-135 cm, tail 3 cm
Distribution: most of Britain and Europe, but not Ireland.
Observation: throughout the year.

The Roe Deer is found in places where there are woods or copses for shelter, or in large expanses of fields that are sufficiently vast for danger (such as humans or dogs) to be apparent from afar. It is originally a forest animal, which has adapted to the fragmentation of this habitat and to the appearance of crops. It is mainly crepuscular and spends part of the night and day ruminating. Its eyes reflect brightly in torchlight. When disturbed it gives a series of loud barks, either in quick succession or more spaced out. They are deep, hoarse barks, which can be confused from afar with a fox or lynx, particularly when echoes and obstacles deform them. They are much more frequent than those of the Red Deer, which are always isolated and more resonant.

Alpine Ibex *Capra ibex*

Length: head and body 105-150 cm, tail 12-15 cm
Distribution: Alps.
Observation: throughout the year.

In summer the ibex spends the main part of the day ruminating on a calm, rocky ridge, rising from time to time to graze. At dusk it descends to richer grassy areas, before returning at dawn, although this pattern can be reversed in winter. Ibex often move around at night, unlike the Chamois and Mouflon, which are distinctly more diurnal.

Index

Bold characters refer to illustrations